Horseshoes and Fences

What doesn't destroy you, makes you stronger!

David Kentish

All rights reserved. No part of this book may be used or reproduced in any manner whatsoever without written permission except in the case of brief quotations embodied in critical articles or reviews.
This book may not be lent or re-sold without permission of the Copyright holder.

The Copyright © is retained by the author, David Kentish of P O Box 359, Goomalling Western Australia 6460.
dkentish@westnet.com.au
www.davidkentish.com.au

Book and Cover designed by
David Kentish
Hand-drawn illustrations by Kaylee Thompson.
All rights reserved July 2020

ISBN 978-0-6487149-8-9

Contents

Prologue	vii
The First Block.	1
Our New Windmill	14
The Loose Boot	37
The Wooden Wagon with the Wonky Wheel	51
When The Farm Went Metric.	62
The fox and the wire	69
Fencing	80
The First Tractor	94
The First Truck	107
Going Bulk	118
Fire	128
Milking Cows	136
The Pigs	148
Modern Stuff	158
About the Author	173
Other books by David Kentish	175

Prologue

In the year of eighteen thirty-eight, my great-great-grandfather brought his wife and family from England to settle in Australia. They had been a farming family back in the "Aulde Country" and they pursued the same profession in the country of their new home. After many generations, the family has spread itself over all states and territories of this magnificent country in their bid to live successfully and provide food for the nation.

Although all components of these short stories that make up this tome are correct, I may have taken some literary license with some of the finer details so that you may be more entertained.

I have taken the path of writing as though the old man is being interviewed by an author who is compiling the details so she may write a biographical story of the life of this Great Australian Farmer.

Because our family has spread so far and wide and been involved in all of the different aspects of agriculture, I have elected to not name a state or town to which the farm is close by.

You will notice that the setting for the farm would be so very similar with many towns or locations within Australia.

Many of these stories have been handed down from the old family and I have had many experiences in the different farming operations that I have been involved in over my lifetime. That personal contact with the land tends to leave one with a desire to keep close to the land and tell some of the stories that happen with those who live on the land.

The chapters are not arranged in any particular order but you will pick out those which are of the early times and those which are more recent.

Not all experiences are funny but when you look back at those situations later in life you can see the amusing side to them. Wherever possible I look at these situations and attempt to provide the lighter side of the story. We all know that life can be very difficult on the land and some light-heartedness does provide some relief.

That is my desire here.

I thank Kaylee Thompson, who is my granddaughter for her work as an aspiring illustrator. This is her first commercial work and she has several projects in hand, so we can expect more from her in the future.

Please read and enjoy!

David Kentish

The First Block.

Well, of course, I wasn't there when it all began because I wasn't born yet. My Granddad bought his first one hundred and sixty-acre block of land that was seven miles from town on the back road after his father had left him some money in his will. Apparently, his father or my great-grandfather was a successful farmer closer to town but he died when his horse and buggy overturned. Granddad's uncle took on the old farm when he passed away and Granddad worked for him for a few years before he took up his own land.

The block that Granddad bought was all covered in bush and scrub with many large red-gum trees as well. He told me that there were about ten trees to the acre. The money that he had, covered the cost of the block, a pair of good workhorses as well as a riding hack. He also spent some money buying tack and gear for the horses.

One of the first things that he built was a shelter for himself and the horses as the first year on the block were very wet as he arrived in early winter. It rained nearly every day.

He survived the first few years living off the land and he fed himself on kangaroo, possum and wild turkey. He made several attempts to make a vegetable garden but while he was working during the day the 'roos and the newly imported rabbits would feed themselves on the good tucker that he provided. After his third year though he did purchase some netting wire and he fenced off a section of about an acre that would become his vegetable patch. It's still there today and I remember some great stuff being grown there too.

The government department that looks after the farmers, or is supposed to, The Department of Agriculture, told all the farmers to clear the land and grow food for our state and country. As the numbers of the population were increasing, this was out of necessity as the state didn't have the capital to import that amount of food for the population. Or so the farmers were told.

So Granddad did just that, he cleared the land, he planted his crops and in the good years, he also harvested his crops. Some years the rain was so abundant that his crops failed because of waterlogging. Some other years the rain was so infrequent that what he did harvest was only just enough to keep the rats and horses fed, thus leaving no income for that year, either. And when it looked like he was going to have a bumper crop the frost moved in and took its toll on the grain.

With the two horses in harness, he hitched them to a heavy chain and had them walk about ten metres apart dragging the heavy chain between them. This method was used to flatten down the low scrub that filled in the area between the trees. By working in areas of several acres at a time he was able to control the amount of area that he cleared at any one time. His idea was to clear an area and then seed it down for cropping. Once the scrub that was flattened down had dried out, he set fire to it. Mostly it burnt well and the fire didn't travel far from the flattened scrub.

At first, he left the trees behind as he thought that they would provide coverage for the new crops as he didn't know a lot about cropping. Neither, it seemed, did the department. The trees obviously held the soil together and provided shade but it did hinder the farming operation.

Granddad told me how he eventually got rid of the trees. After he had successfully cleared about one hundred acres and had this under crop and he had harvested several good crops he took out most of the trees. He purchased a tree puller. This was a winch that had a long handle and was anchored to the base of one tree. A long cable attached to several double pulley blocks was fed out and attached well up the trunk of the tree to be pulled over. When the chains were attached he would swing on the handle of the winch. It had a good ratchet on it so he could take a break at any time. As the cable tightened so the tree began to move. With more tension on the winch, the tree eventually fell over. It was just as well that he had a long cable on the tree puller or the tree would have landed right on top of him. He found it a good idea to have an escape plan if the cable should break. He reckoned that he'd have several seconds to jump clear if that happened. He really couldn't see himself working the farm after being injured by a flying cable.

Then with an axe, he would cut off all of the branches that he could and left the tree to dry out. He continued this process until he had several acres of trees lying on the ground.

He sold some of the timber to the local sawmill and they sent out a team of men with some heavy gear with a team of work-horses and dragged the logs off the farm and into town, to where the sawmill was situated. The sale of this timber kept him going for several years while he was clearing the rest of the block and planting and harvesting his crops. The trees that the 'mill didn't take, he burnt on the ground after the first rains of each season before crop planting began. He'd get his horses harnessed up and with snig-chains, he pulled the logs together so they would burn. Sometimes he needed to use the crowbar just to get the logs close so that they would burn properly.

The branches he heaped up onto the top of the pile and with many weeks of days and evenings of back-breaking work and the stoking of fires, he had his job done. It was very heavy manual work and by day's end he was very tired so he did sleep quite well in his younger days.

After seven years he had the one hundred and sixty acres in full production.

After he had burnt the scrub he'd walk over the area with a mattock, making sure that when he swung the mattock it was well clear of his boot. No good getting blood on the tools, it only made them go rusty and he might get a sore foot into the bargain. Any root that was poking its head above the ground would get grubbed out. He made a sled out of some bush timber and this was dragged along behind the horses. He'd throw the roots onto the sled and when it was full he'd take this over to one of the heaps and unload the roots onto one of the piles of timber that were ready for burning.

The next job to do was to plough the land that he had prepared for the crop to be planted. He purchased a single furrow mouldboard plough that was pulled by his two workhorses in harness. This plough was much too heavy for him to lift so the horses had to drag it everywhere that Granddad wanted it to go. He'd just fling it onto its side and they'd drag it to the beginning of the area that he wanted to plough. Then he would take hold of the handles and balance the plough on the point of the plough-shear. As the horses moved forward, the point would dig into the ground and the mouldboard would turn over the sod. It became very noticeable to him just how important it was to have all of the roots removed from the ground. If the point of the plough hit a solid root, the handles would rise up into the air very rapidly, lifting him well off the ground. He'd have to stop the horses very quickly or else he'd fly over the plough and land on the rump of a horse, or even worse on the ground in front of the approaching plough with no one driving the horses.

This was a very effective way of turning over the soil in readiness for planting the seed. Usually, he would mark out a square area for ploughing and when he arrived at the far end of the row he would drop the handles of the plough to the ground and it would disengage itself from the ground and the horses would drag it to where he would begin the next row. He continued this up and down the rows until the section was ploughed.

The grain was sewn by hand in the early days. He made a bag that had a strap over his shoulder. The bag was filled with seed and he would walk along the area that he had ploughed and broadcast the seed by hand until he had covered the area. Once this was done, he hitched one of the horses to the set of harrows that he had and by running over the seeded area he could cover the seed with soil. It was all very hard and heavy work but he seemed to thrive on it and he usually seemed to grow very good crops when the seasons were favourable.

So with the crop in the ground now he had to wait for the rain to fall at the right time so the crop would grow and mature, set heads and be ready for harvesting in the early summer.

In his first years, the crop was cut by hand with a sickle and a stalk wrapped around a good handful to make a sheaf. This was left lying on the ground to fully dry out. At the end of the day, he would walk around the harvested area, collecting about a dozen of the sheaves and by throwing the cut ends of them to the ground he was able to have them stand up, supporting each other and form a stook. The stook looked like a pyramid when it was built properly with the heads of grain at the top. These stooks would allow the seed-heads to dry, ripen and harden up in readiness for loading onto his sled and taking back to the bough shed he had made. The grain was threshed out of the heads by bashing them against the ground and then sweeping up the grain. This produced a lot of chaff that he kept for horse feed and some of the stalks for the mattress for his bed.

It was a very laborious and time-consuming job.

The grain was shovelled into bags and sewn up. When he had 50 bags he got his neighbour to cart them to the railway siding for loading onto the train that would take his bagged crop to the market near the city.

He had been on the farm for ten years before any sort of mechanisation was available. The first machine he purchased was a thresher. The crop was still harvested by hand (he had some help now) and the stooks were still carted to the threshing machine that was operated by a single horse, through a mechanical rotary device. It certainly took a lot of the back-breaking work out of the harvest.

Several years later he watched his neighbours who had purchased a binder. This machine was drawn by two horses and a cutting bar would cut the crop off low to the ground. This then fell onto a belt that took the stalks to a section on the machine that would form sheaves and tie them with string. When he saw how successful this machine was he purchased one for himself with a bit of help from the local bank.

About this time he built himself a wagon to cart the sheaves from the paddock to the thresher and the bagged grain to the railway siding. He also needed more horses so he purchased several Clydesdale mares as well as a stallion and began his own breeding program.

All the time between seeding and harvesting he was continually buying more land and clearing it ready for cropping.

By the late 1800's he had acquired one thousand acres. His old single furrow plough was too small so he invested in the new CH Smith stump-jump plough. This needed a team of six horses but covered eight times the ground for the same effort. Then with more banks' support, he purchased one of the new horse-drawn Ridley strippers that would harvest the grain on the ripened crop and fill the bags from the hopper.

The horses were kept very busy during those years.
Gradually his efforts paid off and even after several bad years of drought and crop failure, he struggled but was able to pay off the banks and proceeded to operate his business on his own capital.

During this time, of course, he took the time to find himself a wife and made her a new home complete with an iron roof and hard-packed clay floor. It was one of the most modern dwellings in the district. As he was so handy with his hands he made the majority of the furniture from bush timber, and there was always plenty of that. The legs of the kitchen table were buried in the floor to keep it rigid. The same deal with the bed. The bed was something which came a bit later. They needed their sleeping arrangements a little further from the mice, ants and bugs that always seemed to find their way into sharing the bed with them. There was the occasional snake but that's another story.

As the farm grew so did the family and soon his sons, my father and uncle, were helping on the farm that he had worked so tirelessly to establish.

Our New Windmill

Ah, yes the new windmill. Well, it actually was the first windmill in the district too but that was a long time ago. But before we had the windmill though, I do remember what led up to us buying the windmill.

I remember helping my Dad dig one of the first wells for water in the district.
I was about 15 at the time so that makes it such a long time ago.

Dad had seen a couple of blokes dig a well when he visited his cousin down by the coast and reckoned that the two of us could just about make a well so we could draw the water that we needed for the house, the horses and the milking cow.

In the past, we used to rely on the tank that was filled from the roof of the house. The roof was covered in corrugated iron and Dad had fitted gutters to the bottom end of the roof and then put in some pipes to carry the water to the tank. The tank was near the kitchen so that Mum didn't have far to go when it was bath night. When the tank was empty we would hitch the horses to the wagon and with some empty drums head off into town and draw some water from the government well that was nearby. It used to take us all day to cart enough water for us for the week and it was hard work.

After pulling up at the tank by the government well, we had to bucket the water from the open-topped tank and fill our drums. I think the drums came from the road-building contractors and were originally filled with tar for road building in the town. We had 20 of these 44-gallon drums. As 880 gallons is nearly a ton of water, the return trip was slow and we always had to stop several times so the horses could blow for a while.

The stable and the feed shed had a roof of thatch and they were not suitable to collect rainwater. We used to replace the thatch every couple of years as it fell apart fairly often.

Dad reckoned that as the roofing iron was so expensive, that the well would be a cheaper option.

He was always looking to save a few quid!

He got hold of some six by two hardwood timber. This would be used to line the well to stop the soft soil near the surface from falling down on us as we dig deeper. We cut the ends so that they would lock together and the pressure of the soil would hold it all firm without the need to spend money on nails or bolts.

Dad had seen a bloke, I think he was a mate of his cousin, divining for water, so he reckoned that if the other bloke could do it, then, so could he. He took a lightweight green twig that was in a vee shape and held it in his hands and walked for hours over the place. He walked up to the top paddock and down to the bottom paddock and across to the north end of the place but he couldn't get the stick to move. He did it all again the next day too, holding the stick a bit differently but even that did not get him any better results.

While Dad was spending his several days doing his "divining", Mum looked in an old magazine that she had and found a picture of a man holding a stick while he was divining for water. She reckoned that Dad was not holding it the right way. So when he came in at tea-time she told him. Of course, he would have none of this rubbish, he had seen this other bloke do it and he was doing it the same way. But by lunch-time on the third day, he gave in and tried the way that Mum suggested. To his surprise, it worked. The stick was moving so much in his hands, that all of the bark was stripped from it. His hands didn't get sore because they were so tough from all of the hard work. Not like mine are now, all soft with no calluses left. He found enough water to fill all of the creeks, tanks and dams in the district, or so he thought. Then it took him another two days to select what he reckoned was the best place for us to dig the well. It was just past the stables, where he reckoned was the best pull on the stick. It shouldn't be far down, he said, probably about thirty feet he reckoned.

So after heat treating the pick to harden the point and shortening the handle on the shovel, we started work.

Now if you've dug a well and swung a pick and shovel for eight or ten hours a day you'd know that it's hard work. It was even harder work than a day carting the grain from the paddocks and even loading the bags onto the wagon by hand.

It was bloody hard work!

We dug the hole which needed to measure four foot six by three foot six, down to six feet in the first day. That was Dad on the pick with me on the shovel, then me on the pick with Dad on the shovel for two hours about, "so we didn't get bored". From this depth onwards we would need to use a bucket and rope to haul the mullock from the hole. After the first day on the rope with me down the hole with both the pick and the shovel, Dad reckoned that we needed to make a windlass to wind the rope up in to lift the bucket.

We knocked off early that day.

Next morning Dad had this idea of how to make the windlass. He dug a post-hole either side of the well, about two feet away and cut a vee notch in the top of two large posts that we put into the holes and packed them in tight. A round post was laid across the top of the two vertical posts. Then a crank handle was fitted into one end of the round post and this was used to turn it. That was hard work too when the bucket was full on the way up. So we modified the ends of the round post by drilling a hole and fitting a heavy pin into it and then making a bearing for it to run in. Now it works well. A bit of sheep's fat was a good lubricant and made the job a bit easier which Dad seemed to like.

So we worked away. I'm down the hole for two hours on the pick and shovel while Dad swings on the windlass handle, lifting the full buckets of soil and after emptying the bucket, letting it back down again. After the two hours were up, we would swap jobs.

It seemed funny to me but when Dad was down the hole, the ground always seemed harder as he just couldn't get as far as I could when I was down the hole.

To stop the loose stuff from falling back down the hole, we fitted our timbers to the top six feet and took them to about two feet above the ground. We needed to empty the bucket far from the well so that the mullock wouldn't fall back down again. Dad was not too good at that and every so often I got a hit on the head with a rock or some loose dirt.

After several days of this hard work, it seemed to get a bit easier, even with the ladder that we made out of bush timber so we could climb up and down. Perhaps we were getting fitter but I could still dig further in my two hours than Dad could. By the end of the fourth day of digging, we had hit thirty feet and there was no sign of water. The ground did show some moisture and the sides held up well, but there was still no water.

I remember Dad saying about then, "never mind son, we'll hit the water tomorrow".

Well, we kept on digging for ten hours a day and by the end of the week, we needed a break. We always have Sunday off so this was a time when we could relax and just think about the well.

Monday morning came and we spliced on some more rope so we could reach the bottom. By Monday night we reached forty feet and there was still no sign of water or change in the soil type. We kept on digging Tuesday, Wednesday and Thursday. By Friday morning there was still no sign of water but Dad was adamant that this was the best place to dig the well so we kept on digging, taking two hours about, "so we didn't get bored". When we knocked off Saturday night we had reached fifty-five feet and there was still no sign of water.

We were all getting a bit despondent by this time and when we went to church in the morning, several of our neighbours asked us how the new well was going. Well, everyone had their own idea of how to dig the well but none of them had ever dug one so we just didn't listen to them too much. The most common comments were, *"there's no water around here, hasn't been for a hundred years,... you should have dug the well down by the creek,(that was outside our property),.... You need a professional well-digger, now my mate in the city, he knows a bloke....* Well, we left them to their comments and we headed off home again.

Next morning we were a bit slow at getting back to work so we took longer to feed the cow and horses and carry their water from the tank which by now was nearly bone empty. We eventually got back to the well after smoko that Mum had brought down for us, knowing how we were feeling. We sat down by the well and had the cuppa. Dad picked up a stone and tossed it at a crow that was sitting on the windlass but he missed the crow and hit the barrel of the windlass and the stone fell down beside the well, so he picked up another stone and threw it at the crow. This time he hit it and off it flew to the stable roof to peck at the straw on the roof. When the crow was about half-way to the stable there was a splashing sound. We all looked at each other. There's no water here to make a splashing sound, the well was dry last Saturday when we finished work.

I wandered over to the well and looked down and there was this little patch of shiny stuff looking back at me. It was like a mirror. Yep, sure enough, there was water in the well! Well, we all laughed and Mum and Dad linked arms and danced a jig, then Mum grabbed my arm and we danced a jig. Dad grabbed my arm and we did a jig too! After the initial excitement had abated, I climbed down the ladder and stood in well with water up to my knees.
I yelled back up to the top of the well and told them how much water there was and Mum soon worked out that over the weekend we now have

about 200 gallons of water. Dad sent down the bucket and I filled it with the water which tasted sweet and he hauled it to the top. Both he and Mum tasted the water and were happy that it was good enough for the kitchen. Dad fetched a couple of drums and we continued to bail out the well and fill the drums. It didn't take long for us to reach the bottom. I could see that the water was just seeping in but we needed to get deeper if we wanted more water. So I kept on digging in the mud. Because the soil was wet I was able to dig much quicker but I still needed to use the pick. The water kept on coming in but I was able to keep up with it and still dig out more soil.

We made ten feet more that day and that took the well to sixty-five feet to the bottom. I stayed in the well for an hour after I stopped digging and in this time the water had risen to my knees again, so we reckoned that that should give us enough water for us and the stock.

Because the wagon was very high, Dad made up a sled and we fitted a couple of drums to it and tied them down. This was parked by the well and using the windlass and bucket we could draw water from the well and fill the drums. The horses dragged the sled over to their water trough and I bucketed the water from the drums to the trough each day for the horses and the cow. When the horse trough was full we'd take another load of water and leave it parked by the house, so Mum could draw water from the drums for the kitchen.

This was the practice for many years.

The windmill didn't come for another twenty years or so.

After Mum and Dad had passed away, my wife and I moved into the old house. I was still carting water but had to replace the bearings on the windlass as they had worn out. I think I made a new bucket and fitted a new rope too.

Our boys were attending the same school as I used to and were approaching their final years and soon they would be on the farm full time to help me.

One of the boys was a good reader and saw in an American magazine some information about a windmill that was used to pump water from a well and deliver this water into a tank. Apparently, they have had these for some time and they worked well. When I was in town the next time I called into the local stock and station agent and asked him about the windmills. Yes! He has heard of them and would chase up on some details for me.

A couple of weeks later he came out to the farm on his horse and brought with him some printed details of the Aermotor windmill that was made in a place called Chicago, way over in North America. It was a bit pricey but after the good harvest, we had last year I decided that we could afford the cost of the windmill with its pump and pipes but we would do the work ourselves, the boys and me.

Several months passed and the agent saw us in town one day and told us that the windmill would be at the railway station next week. So the next week, the boys and I took the wagon into town to pick up the 'mill. The station master took us to the rail wagon that held our 'mill and our faces dropped as we saw that it was all in pieces and a couple of boxes. Anyway, we loaded all of them onto the wagon and took it back into town and to the stock and station agent. "Oh, Yes!" he said it comes dismantled to save on cartage costs but it's easy to assemble as there are instructions packed in one of the boxes. He helped us open the wooden crate where we found the instructions and the boys sat down to read them. (They were faster readers than me.)

The boys said that we'd need a few spanners, some cement, some pipes and pipe tools.
My goodness! The cost is getting bigger all the time.

The agent had what we needed, so we were soon on the trip back home again with everything on the back of the wagon. It was late when we arrived home, so after un-harnessing the horses, feeding and brushing them we headed into the house for tea. We'd work out how to build this windmill thing in the morning.

To my surprise, the boys were up before me. This was unusual but great to see.

After our early breakfast, we headed out to the wagon and set about making a windmill out of all of the bits and pieces that we had brought back from town yesterday on the wagon.

There were two large boxes and several bundles of steel angles all strapped together. The boys had read the information and found out that the bundles are marked so they have a bottom section, a middle section and a top section in the bundles. They found the bottom bundle and unpacked it and assembled it on the ground according to the instructions. It went together well and next they bolted the middle section onto the bottom section. Next, they unpacked the top section and bolted them all together to make a 30-foot tower.

The instructions indicated that now is the time to dig the holes for the bottom bits to be fitted into. The boys tell me that these are called the "anchor posts" and are put into some holes in the ground and filled with concrete. Of course, these holes have to be in the correct place. The boys soon worked this out and began digging.

By the time we finished work for the night, the holes were at the correct depth and had a footing of concrete at the bottom for the windmill tower to sit on so it will be perpendicular. Apparently, this is important for the mechanism to function correctly. The instructions said to leave the concrete to set for a day or two so that it will be cured. So we did.

The boys weren't ones to sit around, so they got to work sorting out the bits and pieces that they needed for the top or head of the 'mill. The wheel segments or sections needed to be bolted together in the correct order so that they will make a complete wheel that is balanced when it's fitted onto the head at the top of the tower. The boys proceeded with this and by the end of the day, there were not many parts lying around that needed to be put together.

Next morning the boys tell me that we can now stand up the windmill tower. The book shows several ways to do this and after an hour of discussion, we decide to shift the tower so it's in the right position near the well and the holes that we have dug. A long post is stood up with a rope over the top of it to the top of the tower. This rope is tied on firmly and the other end is attached to the chain ready for the horses. Two more ropes are attached to some posts that are fitted into some more holes that the boys dug on the other side of the tower. These ropes are attached to the top of the tower after being carefully measured. They will stop the tower from falling over when the horses pull in the other direction. The horses are attached and I move them forward slowly as the slack in the rope is taken up. The horses put their shoulders into the strain and slowly the tower raises off the ground using the fulcrum of the long post to lift the top end of the tower up. One of the boys takes care of the long post as it comes free as the rope extends out from the top of the tower. We didn't want it to fall over and hurt someone. The tower continues on its travel to the apex of the arc and it rests neatly in the holes that have been made ready for it.

Now that that part is done, it's time to mix the cement to make the concrete for the legs of the mill to be fixed in. We took two horses and the dray down to the creek that we drive past on the way to town and get some river sand and stone for the concrete. This we shovel onto the dray, being careful not to overload it so that the horses can pull it out of the creek bed. When we get back again we begin mixing the cement on the ground. Back then we didn't have a cement mixer and all of our cement had to be mixed by hand on the ground. My Dad showed me how to do this many years ago when we cemented in the back veranda of the house. After several hours of mixing cement and filling it into the holes, the job is done. The boys checked with a spirit level just to make sure it was all perpendicular. We made one small alteration and then we were all happy. After cleaning up the tools it's time to knock off for the night as it getting dark and Mum will have tea ready for us.

The next day is Sunday and we take two horses on the jinker and off we go to church. There is a lot of talk about the new windmill that we are putting up. Apparently, the agent has boasted to everyone about his sale of the windmill. We told them how well it was coming together and by next week we could tell everyone how well it all works.

On the Monday morning, the boys and I start putting the top onto the windmill. One of the pipes that we are to use down the well is stood up beside where the "engine" goes at the top of the tower and is fixed into position with a piece of timber and some "u" bolts. A pulley is fitted to the top of this pipe and a rope is passed through so both ends are at the ground. One end is tied to the engine as per the instruction and the boys decide which of them would go up to the top and work up there. We hauled on the rope and pulled the engine to the top and it fitted over the pivot tube and lowered into place. Next, we sent up the tail and it was fitted and then we tied on the spokes and sent them up. After they were bolted on, we sent up each section of the wheel. They were fitted on one at a time and then both boys worked up the top just to finish off the wheel and fill the gearbox with oil. When the boys were happy that everything was finished by the book they tied the wheel off to the tail.

The next morning we began the work on the pump and pipes. The pump is a brass cylinder that had a plunger with a rod sticking out the top of it. Each of the pipe sections has a rod inside it. This had to be coupled to the rod in the next pipe to join them from the windmill engine to the pump. When the boys got this done they connected another pipe to a "tee" piece to deliver the water away from the windmill.

The last job to do was to untie the wheel so that it could face the wind. There was just a gentle puff of wind and it began turning. The pump-rod went up and down as the wheel turned and not before too long, the water began running out the spill-pipe.

Well, I remember that we spent quite some time rejoicing with excitement as everything now works. The boys were jumping up and down and clapping and yelling with giant grins on their faces. It was the happiest I had seen them for a long time and that memory will stay with me forever. Mum heard all of the noise and came to join in the excitement. We all jumped around and hugged each other in our excitement.

After a time there was quite a puddle on the ground under the end of the spill pipe and we used the horses to haul the sled under the flow of water to catch it. When the drums were full, we turned off the 'mill so we didn't waste the water. The sled was dragged over to the horse trough and it was filled and then the rest was taken to the house for Mum to use.

We could see what we had achieved with this new windmill but it just wasn't quite finished.

We needed to put in some pipes to connect the windmill to the tank.

There was still enough left in the bank account for us to buy the pipe-work for that purpose, so the boys and I took the wagon into town and picked up some more piping that we could use to finish the job. After loading the pipes and a bag of fittings we headed back home again. During the next day, we soon had the pipe-work laid out and by that night we had the windmill pumping directly to the house tank and the horse trough.

It was such a pleasure to sit outside in the evenings enjoying a quiet drink and watching the windmill work, pumping the water that we needed for the house and the horses and the milking cow. Now we didn't need to waste all of that time taking the horses into town and carting water back to the farm.

We can find something more enterprising to fill in our time.

Yes, we did tell everyone when we were at church the next Sunday and we were very happy to have the windmill. It made so much difference to our farm. It was about the most convenient advancement that we had made back in those days.

The Loose Boot.

You know, on the farm we had many sorts of boots.

Of course, there are those which are worn on the feet and even in this instance the types of footwear also changes with the season or the event.

How so, you might ask?

Well just think of this. Whilst we are working around the farm and the ground and weather are dry we would normally wear leather boots. These used to have a leather sole with studs in them but in the late 1930's they developed a rubber sole. The early ones didn't last long as they kept falling apart but they did become improved as time went by. Sometimes these would be the elastic-sided boots but they didn't arrive into our use until the 1970s. Before that, we had to use the lace-up boot and they could take a few minutes to do up and then take off at the end of the day. On the other hand, they did tend to last longer as the modern elastic-sided boots lose their elasticity after a few months of hard work. They do become loose.

Then conversely when the ground and weather are wet we would need to wear our rubber boots to keep our feet and lower legs dry. We also would wear these mostly in the dairy as there is always water flowing around. They also come in very handy whilst we are slaughtering animals as these are easier to clean. Just hose them off in fact. They are also great whilst walking through the wet grass. Those leather boots can become very soggy when they are wet and they take quite a while to dry out too! I remember one day coming in from the paddock after a summer thunder-storm with wet boots so I tried to dry them off quickly in the oven in Mum's kitchen. The oven door was open and they laid on the open door with the heat of the oven wafting out over them. I thought that they would dry quickly while I had afternoon tea and be ready for me when I went out again. Imagine my amazement when I went to take them from the old oven door, they had shrunk so small that I couldn't get them on again and they were so hard! I tried all sorts of tricks to get them back to their original size but all to no avail. They were useless after that! Luckily they were nearly worn out and I had the new pair in the cupboard ready for me to use.

The other sort of boots which I used is, of course, the dress boots that I'd wear when we go to town or the dance. Well-polished, all shiny and black.

I think that takes care of the footwear part of boots.

Another sort of boots that we use on the farm are the boots fitted to the tynes of the combine. You know! That machine that we use whilst planting our crop. These are metal boots that are shaped a bit like a funnel and are riveted or bolted to the tyne, just above the point which is what does the job of breaking the ground. Seed and fertilizer falls through the dropper tube from the seed or fertilizer box by way of a flexible dropper tube. This tube fits loosely into the boot which allows for the movement of the tyne arm as the job of seeding progresses.

Another sort of boot that's on the farm is the one at the back of the car. The boot usually carries a spare wheel, well, that is if I don't forget to put it back after I have repaired the puncture. There are also a few basic tools and a jack if it has been put back after changing the spare on the ute. The boot of the car is also very handy for when Mum goes shopping. She can load the groceries in the boot after she has lifted in the 40-pound gas cylinder, which she needs for the kitchen cooker.

Another type of boot is the one that is attached by a leather strap to the saddle of the horse. Into this boot usually fits the rifle. Mostly I prefer to have the rifle between my leg and the horse as this keeps it more stable when galloping across the paddock chasing the 'roos and emus that have been destroying my crops. It's fairly easy to remove from the boot and raise to the shoulder for a good freehand shot from horse-back. You just need to make sure that you're on the horse that has been trained for the purpose. I remember once when I was on a new horse which was supposed to have been rifle broken and I shot from the saddle. Needless to say, the 'roo got away and I walked home from the back paddock. The horse was by the hitching rail when I got there. The bruising only took a week or so to heal.

On old truck that I bought, to help with the seeding, had brake trouble and I needed to replace the slave cylinder in the wheel hub. Everything was going well. The opposite wheel was chocked so that the vehicle could not get away from me. The chassis was held up by the old jack and a few blocks of wood just to make it safe. The new slave cylinder was fitted to the backing plate and just before I assembled the shoes I remembered to fit the boot which keeps the dust from the cylinder. The funny part was that I never did find that spring which I was trying so hard to fit onto the shoe to hold it in place. That spring shot off over my left shoulder at great speed when the pliers slipped. I always used a short piece of light wire for the purpose after that experience. Huh, I do remember that so well. It could have been much worse if the spring had hit me in the face. I didn't wear glasses then!

So that's another boot.

My uncle who farmed several miles away hired some workers one year to pick up the sticks and roots that had come to the surface during ploughing. He had the old horse and dray and they were all working alongside each other picking up the sticks and roots and throwing them onto the dray. When the dray was full, uncle would walk the horse over to the heap they had made and tip the dray to unload it. The others would take a break while he was doing this. When he returned they carried on as they had before, throwing stuff onto the dray. One of the blokes must have been a bit over-exuberant with his swinging action of a rather large piece of wood because it sailed right over the dray and hit uncle on the back of the neck. As he didn't see anything, just felt the sudden pain, he kept his stature and continued on without so much as a flinch. He thought that someone was playing a joke on one of the others and didn't think much more about it. Several minutes later the same thing happened again. His reaction was just the same. Anybody watching would have thought that he had no feeling as he just kept going as if nothing had happened. A little later, aunty came along in the buggy and brought out smoko as she did every day that they were working at clearing up after the plough. She just happened to see one incident but kept quiet about it for a while. When they were halfway through smoko she quietly asked uncle about it and he mentioned to her that it had happened four times before that. She asked him what he was going to do about it. He told her that when it happened again he would find the bloke who did these things to him and give him the boot. Before lunch, which aunty had left for them, uncle felt another blow to the head. He immediately spun around and faced the culprit, who was stunned to see this reaction that he had been waiting for since his second throw. "Sorry boss that one just slipped out of my hand."

"Sorry be damned!" uncle thundered, "You've tried to kill me five times and each time I turned the other cheek. I've had enough of your insolence and am giving you the boot!"

"What's the boot for? I've got some good ones here already," Replied the worker, sheepishly pointing to his old boots.

"No, you fool, you've got the boot, you are sacked, fired, kaput! Finish immediately!"

So that's another boot that we deal with from time to time.

But the boot which causes us the most problem is one I touched on before and that is the foot-wear type. As I mentioned before those leather boots tend to swell and get very soft when they get wet. Well, this caused some problems for me one day when I was out on the horse checking the stock. I was about an hour from home when we had this heavy drenching thunderstorm and I got saturated. So did the horse. There was so much rain, about an inch I think, that the creek came down in a sudden rush and a cow got stuck in the mud beside the creek. Well, I had to get her out or she would have drowned, with the creek rising as it was. I tied a rope to the saddle and onto the cow's horns and backed up the horse to keep a tight strain on the cow. Then I got off the horse and waded into the mud and pushed her from behind. I slipped. The mud wasn't very deep but it was quite sticky.

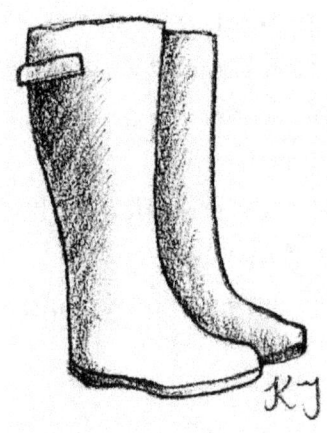

Soon the cow began to get the message that we were trying to help her out of a difficult situation and gave us all the help she could. It wasn't much but it did help. After slipping into the mud a few more times and vocalizing to the cow and the horse, who did seem to notice that I was getting concerned, we proceeded to dislodge the cow from her predicament. After I untied the rope from her horns she put her head down and took a run at me.

How rude!

After all the work that the horse and I had done to free her and that was the gratitude she showed us. Fortunately, she only took several paces before she shied away and took off across the paddock looking for the rest of the herd.

I was covered in mud.

The horse was covered in mud but I climbed into the saddle. The boots were so slippery and swollen by the water and mud that they were a tight fit into the stirrup irons. We had several gates to go through to get home and with the old post and wire "cockie's gates" I had to dismount to open and close them. The thunder was still banging away and the horse was very skittish but I held him.

For a while.

At the second gate, the thunder crashed as the lightning flashed not 100 yards away. The horse shied and took off just as I mounted and I didn't get my right boot into the stirrup iron. The horse then shied at a wet possum that ran across the track and I went sailing over the other side of the horse. So now here is the horse galloping along with me caught by the left boot in the stirrup, dragging along beside and slightly behind him.

Gee, I remember those rear hooves of the horse were very close. They seemed to get very big all of a sudden.

I tried to stand so that I could run beside the horse so I could get out of the sticky situation but then I remembered that you need two legs for that and one of mine was sort of out of action for the time being, still stuck in the stirrup iron. So the best I could do was try to keep all of my appendages close to my body so they wouldn't get tangled in the horse's hooves or the sticks and stones which were scattered around. Now that's not quite as easy as it sounds. Bumping along at fifty miles an hour *(well it seemed that fast)* being dragged by the foot by an irate horse is something that I am not likely to forget for a long time.

The horse jumped over a log but he forgot that I was sliding and bouncing along beside him and he didn't have any consideration for my safety whatsoever. My backside collided with the log and the impact lifted me clean off the ground. I think that might have been my saving grace because the boot has swollen with all the moisture and my foot slipped right out of it and I was left lying on the ground after sliding and rolling for several yards. I was shaken, cut and bruised but luckily there were no bones broken this time. The horse just kept on going! He wasn't going to even look at me to see if I needed any help. Oh no! Stuff you jack, I'm okay.

I eventually found my left boot after hobbling along amongst the prickly bush. The walk home must have done me some good because when I found the horse I was too tired to even chastise him. I didn't even rub him down, just took off his saddle, gave him his hay then slipped the rail into place on his yard and went into the house to get cleaned up. Luckily Mum was home to help me.

A few days later I ventured out to find that the horse was fine. But I don't remember ever seeing the cow again. Perhaps she just avoided me.

So that's the story of my loose boot. It probably saved my life. Well, it certainly saved me from more injuries, anyway.

Horseshoes and Fences

Several years ago, Bert, the old farmer from down the road, was driving his old semi-trailer loaded with bulk wheat to the local bins and as I watched him drive past, I was reminded of something that had happened, many, many years before.

I remember that the crop of that particular year was very good and although we only stripped 7 bags to the acre, Bert had a better crop than ours and stripped just over 10 bags to the acre. The crops were planted at the same time with the same variety of seed but he seemed to get the rain just right. Our farms are only 5 miles apart and you really wouldn't think that there could be much difference in the rainfall over that distance but that is what happened.

Back then his two sons helped him load the bagged wheat onto the wagon. They were strapping lads and did most of the bullocking work around the farm. They were 12 or 13 years old at the time and attended the local school when Bert didn't need them to help him on the farm.

Bert had just invested in one of the new-fangled bag lifters that attaches to the side of the wagon and while one lad dragged the bags in and stood them onto the lifter at ground level, the other lad would take the bags as they come off the bag lifter at the top and make a stack of them on the back of the wagon. This stack would go up about six feet high and took quite some work.

To make the bag lifter work, Bert had one of his old horses attached to the lifting arm with a trace chain connected to the horse's harness. When the bag was loaded onto the lifter, Bert would walk the horse forward, this pulled the rope which in turn pulled the arm on the bag lifter and this would lift the bag up and onto the back of the wagon.

Nearly everybody in the district knew of Bert's new bag lifter and some of us went over to his farm and watched this new contraption in operation. We laughed when we first saw it in operation as the horse that Bert was leading, would walk too fast or just a bit too far and the bag would go sailing right over the wagon and land on the ground at the other side. One bag even collected one of his lads on the way over and knocked him to the ground as well. I think it was only his pride that was hurt as he got up and back onto the wagon again while forwarding some very choice words at his father. I do rather think that he was trying to get him to do his part of the job correctly, although it didn't sound like that at the time. After a few more attempts they got the thing working well and it certainly did make the job of lifting those bags of wheat onto the wagon a whole lot easier.

The six-horse team pulled the old wooden wagon with the wonky wheel quite easily on the flat ground and the slight hill between here and the railway siding slowed them down only just a little. They were magnificent looking

Clydesdales that he had bred himself, from some older stock that had been imported from the old country about 30 years before. He bought them from one of those city type farmers who tried farming but just couldn't seem to get things right. He wouldn't listen to us locals who had been brought up in the district and after several years, his capital had expired and before leaving the district he sold off what he could and Bert took on his horses. I don't think the other chap was very happy with the price that they were sold for but there was only one bidder for the horses at his clearing sale and he needed to sell quickly as he was in a hurry to go south.

The two shafters were geldings and in front of them were a pair of young mares with the two leaders being a couple of older mares that had a lot of experience in wagon and plough work. Bert found that the two older mares in front as leaders had a tempering effect on the two younger mares who were always trying to go too fast. Bert has the same team on the plough and when he's harrowing, he used just the two younger mares as they could step out and get the job done just a bit quicker. The stallion was a bit cranky and Bert only used him on the single furrow mouldboard plough when he was planting corn. The old stallion used to bite Bert when Bert tried to put the harness on him but Bert taught the stallion a few lessons with a knotted rope and he didn't seem to bite him any more after that. I think Bert's patience had been sorely tried as he is usually very good at handling cattle, sheep and horses but a man can take only just so much.

The wagon never used to have a wonky rear off-side wheel but Bert was always trying to find ways to save money, so he tried to fix the loose, rattling iron tyre himself. He has a forge where he learned to do some blacksmith work from his father who was a better smithy than he was a farmer and Bert reckoned that he could fix the loose tyre by himself. So after raising the wheel from the ground and supporting the axle on a block of wood, he knocked out the retaining pin and took the wheel off the axle and laid it flat on the ground. The wheel, of course, was fairly large. It stood just as tall as Bert with the iron tyre being about four inches wide. The wheel was also very heavy, but Bert being a big man could handle it without too much strain. He also used levers and blocks of wood, whenever the need arises, to help with the lifting.

After setting out some more blocks of wood on the ground to raise the wheel from the ground he proceeded to knock the tyre off the wheel with his sledgehammer. After half an hour's work, it came loose and fell to the ground just missing his boot. He already had one toe missing from a previous accident and he was careful not to get his booted foot under that falling iron tyre again. He couldn't afford to lose another toe.

Bert proceeded to lift the tyre onto the forge and after a lot of bellows work and adding some more charcoal, he eventually got the tyre red hot. At this stage, he shifted the red hot iron tyre to the tyre shrinker. This he achieved without too much difficulty as he had some good lifting gear on the blacksmith shop just for this purpose. The tyre was shrunk about half an inch in the tyre shrinker and then he shifted it back into the forge and proceeded to get it back to a red hot condition again.

The plan from here was to position the tyre over the wooden wheel which had been left out in the sun to dry and shrink as much as possible and drive it into the correct position using his favourite sledgehammer again. As the iron tyre cooled, it would shrink just enough to fit tightly on the rim of the wooden wheel and he would have a good straight wheel without a rattling tyre.

But like all plans, they sometimes just don't go right.

After Bert had lifted the red hot tyre with his lifting apparatus, he came to grief as he was pushing the tyre over the wheel. The block of wood that was supporting the wheel by the spokes had broken in half (well, it was an old block and fairly rotten in the middle) and allowed that side of the rim to fall to the ground. When the wheel fell down it collided with one of the levers that Bert was using to shift the heavy tyre around and this lever spun up in the air and as it came back down to the ground again it

landed right in front of his foot as he was stepping forward and it tripped him up. As he was falling to the ground he knew that he must not land on the tyre as it was still red hot and that would do him an injury, so he twisted his body away from the tyre and the wheel, landing fairly safely off to one side but not out of danger, letting go of the rope that controlled the lifting apparatus in the process. He hit his head on the sledgehammer which was laying there ready for him to use to drive the hot tyre into place on the wheel's rim. This collision with the sledgehammer rendered him unconscious and he lay on the ground just a few inches away from the red hot iron tyre.

With the lifting apparatus in freefall, the tyre dropped down, just in the correct position and with its own momentum, fitted itself right onto the rim of the wheel. But now the wheel wasn't being supported in the correct fashion and was now buckled. The iron tyre shrunk as it cooled and it became so tight that it was not able to be moved again on that rim and so that is the way it stayed.

After a while, Bert came too and when he had shaken himself free of the dust that had attached itself to his clothes, he stood up and gathered his thoughts. He looked at the way in which the iron tyre had fitted itself onto the rim of the wooden wagon wheel. Although it was fitting very tightly it was buckled. He tried to remove the tyre again so that he could correct the buckle in the wheel to allow it to run straight again but no matter how hard he tried to drive the tyre with the sledgehammer, it just would not budge. It was very late in the day by this time and Bert went in for his evening meal with the family and slept on the problem all night.

Next morning he came out to have another look at the buckled wheel and got to thinking that if the tyre is now tight and not making that rattling sound each time it rotated, he might as well leave it like it is.

So he stood the wheel up and after fitting it back onto the freshly greased axle of the wooden wagon again, he rotated the wheel just to see how far out of true it really was.

Well, it looked atrocious!

It wobbled about three inches out of true and he was not very pleased with himself at his efforts of fixing the loose tyre. He stood looking at the wobbly wheel for an hour, deciding just what he was going to do with it. In the end, he decided that he would leave it like it was and see how well it performed on the wagon. Harvest was to start in a few weeks and he really didn't have the time to have another go at the stupid thing anyway. He jacked up the axle and removed the block which had been supporting it and lowered the wheel to the ground once he had refitted the locking pin.

Bert harnessed the team and soon he was moving the wagon forward. He extended the rains so he could walk behind the wagon while it moved forward and he could see just how bad it looked. Well, he thought, "It is a bit out of line and does wobble a bit but it doesn't rattle anymore, so at least I've fixed that problem."

So that is how Bert's wooden wagon got its wonky wheel.

He carted all of his wheat to the railway siding that year as well as our four loads. For the next five years, Bert carted wheat to the siding with his wooden wagon with the wonky wheel. It amused us after a shower of rain and saw the tracks in the mud of the wagon with only one wonky wheel.

When The Farm Went Metric.

You asked what happened in February of nineteen sixty-six?

This country adopted the metric system! That's what happened!

Up until that time, everyone was using the good old imperial system for measuring. You know, pounds, shillings and pence; acres, roods, perches, square yards and square feet; miles, chain, feet and inches; ounces, pounds, hundredweights and tons; fluid ounces, pints and gallons. That sort of stuff.

It might be alright for the younger ones to comprehend the change from our old imperial measuring system to this new-fangled metric system but for us older ones it was very hard for us to understand and then even harder to

comprehend. So, it was often referred to as "dismal cruncy".
It changed so much of the way that we had to think about things but luckily for me it was the boys who were running the farm by that time.

Do you know that our farm even seemed to change!

I scribbled down some information and stuff the other day as I knew you were going to ask me this question.

Our farm was 7 miles from town but now this distance has increased to 11.2 kilometres.

Over the years we built up the farm by buying up land as it became available from other farmers and in 1966 we had a farm of 16,000 acres. Well, this was reduced to just 3952 hectares. Nowadays, the farm would be 32,000 acres but with the metric system, it's just 7410 hectares.

We used to plant 8,000 acres of wheat each year but after the new system came in we planted just 1976 hectares. We used to buy the 1200 bushels of seed for our planting but now we only need 32.7 tonnes. We used to seed at a rate of 90 pounds per acre and now we need 98 kilograms of seed per hectare.

As for fertiliser, well that's another story altogether. Super was costing us around £3-10-0 a bag and this would be spread over an acre of cropping land. But after the metric system came in, it used to cost us about $17.00 a hectare to fertilise the pasture and even more for the cropping program. You could see why we would not make any profit that year.

And then when it comes to harvest time we used to be able to harvest 30 bushels per acre but now we only get 1.9 tonnes per hectare.
And then when we came to get paid for the wheat that we delivered to the bin we'd get £18-12-6 per bushel but after February of 1966, we got $102.50 per tonne. Just think how much more tax we had to pay!

The transport of the grain to the bin used to cost us more too! We used to pay around 5 bob, that's five shillings, a mile to the carrier but then he wanted 32 cents a kilometre.

Even fuel for the tractors got much dearer too! We used to pay about 4 shillings and tuppence a gallon for our diesel but after the metric system came in we had to pay 10.5 cents per litre.
Talk about a cost increase! And that was way back in 1966!

When it came to tax time we really had our tongues hanging out. Our tax bill for the previous year was £4,500 but now the greedy buggers want $9,000.00.
Where is it all going to end?

The old well that my Dad and I dug when I was a young fella seems to have shrunk. It used to be 65 feet deep but now it's just 19.8 metres. Even the windmill has got smaller too! It used to be 30 feet high but now it's only 9 metres high. The wheel of the windmill used to be 8 feet in diameter but now it's just 2.4 metres. The pipes have changed in size too! We were using a 2-inch pipe but now we have to buy 50 mm pipe just to do the same job. Even the pump-rods have changed from five eighths of an inch to 16 millimetres in diameter.

How bloody confusing it all was!

The tank by the house that feeds water to the kitchen and the bathroom seems to have changed in size as well. It used to hold 12,000 gallons but now it holds 54,000 litres. We used to use 1,000 gallons of water a month for the house but I had to tell the wife to slow down as now she was using 4500 litres a month. So at that rate, the tank would soon be empty because the windmill was only pumping in at the rate of about 40 gallons a day.

Even when the wife used to do the fortnightly shopping in town she complained that the groceries were costing more. She used to spend about £12-10-6 but the next week this had increased to over $25.00. She'd hand over a shilling, that's twelve pence, but only get ten cents for it. For each one pound that she handed over, she'd get two dollars for it. She was just so confused. The milk and bread went up from 4 shillings and 9 pence to 56 cents. Boy what a jump that was!

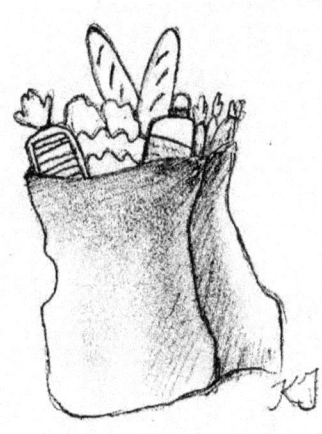

It was just so difficult to get the mind around how everything seems to get smaller and the distances get greater and things cost so much more and become lighter and we pay more tax.

Well, there is one consolation now and that is that it's not my worry any more, as I've retired from the farm work. The boys and their families run the place now and they seem to be doing alright too! They pay me an allowance of $35,000.00 a year for my "consultation" but when you come to think of it that would have been only £17,500!

So it seems that at least I'm better off than I was before.

The fox and the wire.

Oh, thanks for asking that question!

Yes, foxes did play a large part in our sheep breeding program. Even Mum's chooks suffered from the little buggers. Oh, pardon my French.

It seems some of our forefathers didn't think things through all those years ago. They brought in rabbits so they could have a bit of shooting sport. Then they became a problem because they seem to breed, well, like rabbits. Then they brought in foxes to control the rabbits but the foxes preferred other animals for their sport. They hardly touched the rabbits. How dumb could they have been? And the same thing happened when they brought in cane toads to try and control the cane beetle in the sugar industry.

Disastrous!

Foxes have become a big problem and nearly every night we take the ute out and get one or two with the spotlight and point 303 rifles. It's not that we don't like foxes it's just that they seem to kill for the sport of it. New lambs are one of their favourite games. They chase them until they are caught and just kill them and leave the carcasses on the ground. Hardly a bite is taken from the carcase, or they just leave the maimed animal to suffer. If it survives, that is.

If the chook-yard gate is left open at night, they get amongst the chooks and have so much fun killing as many as they can but not eating anything. It's just so exasperating! They cost us thousands of lost dollars every year.

So we get back at them and get as many as we can.

The old paddock ute is a bit battered now and it's just the thing to get around the farm. One of us driving and another one using the rifle with the spotlight attached has become our nightly activity. We usually get to bed by ten, so it's not too late. Sometimes the early start next day at or before sunrise can be a bit difficult if we celebrate too much when we get back though. The old man's a pretty good shot, so that means I get to do most of the driving. Some of those foxes can run too but the old ute can outpace them when it's pushed along. Our best night's shooting was seven foxes in two hours and then we didn't see another one for a week.

It isn't only us who have this affliction with the foxes, as nearly everyone else in the district does the same thing. Nearly every night when there's a light breeze, it almost sounds like a battlefield out there. Little point 22's, some point 243's but mostly the old reliable point 303 ex-army Lee-Enfield rifles are out there. Depending on each person's choice as to which rifle they use. The cost does come into it, as the point 22's are the cheapest at about a penny a shot. The point 303 costs about one shilling while the newer point 243's cost about two bob a shot. Mostly the younger ones use the point 22 until they get their eye in and then they progress up to the larger bore rifles. There are a few shotguns used too, but sometimes you just can't get close enough for a kill with a shotty. There are a few rifles out there with telescopic sights. They do have their advantages but also disadvantages too. If they get bumped and that's a fair chance with some of our paddocks and rough tracks, they lose sight and your shot can be wide, low or high. Now that can cause other problems too, like how far away is the neighbour's house. You really don't want the odd miss to kick up the dust too near their house, or worse.

The mate and I were at a friend's place one Saturday night to play cards but we were not having much luck, so we decided to pull out. The mate got me to drive his ute back to his place and we would go into town to the drive-in to see a movie. He wanted to change his shirt or something on the way. But after what happened later, I think it should have been more than his shirt.

On the way to his place, we were driving along this dirt road, when we saw a skinny old red tail fox run across the road. He must have noticed us behind him as his tail that was pointing to the sky when we first saw him suddenly stuck straight out behind him as he picked up speed. With his head down, following his nose and his tail pointing to where he had come from, it looked like he was trying to get the best out of his aerodynamics. It ran along the road for a quarter of a mile, with us following closely behind, then it skidded sideways and headed in through a gateway into a paddock.

Luckily for us the gate was open so still doing about 50 miles an hour I swung the wheel and the ute broggied off the road and in through the gate a bit sideways but we were able to keep the fox in the headlights for a while. The mate got his rifle and spotty ready and I was to swing to the right of the fox so he could get a clear shot at it out the left window. I swung the wheel of his near new ute and just as he was about to take a shot, the fox just disappeared from sight.

This was lambing season, so we wanted to kill any foxes before they could kill our new-born lambs and we were determined that this one was not going to getaway. But things didn't go quite according to plan.

I swung the wheel in the other direction to control the skid, the headlights picked out a group of trees in the distance. I hadn't been in this paddock before but the mate had, as he crops this paddock each year. The mate who recognised the group of trees, suddenly realised where we were and he screamed out "Oh shit, this is the quarry paddock".

Just as he had finished the "dock" part of his sentence, we ran out of paddock and the ute just dropped.

We both screamed like a couple of girls at a party.

It seemed like we were just in slow motion. Just floating through the air with the engine screaming as the back wheels lost traction.

We fell.

It seemed like a hundred metres down. We were going downwards faster than we were going forwards. Soon our forward motion slowed then stopped and we just dropped, much like a school bag.

The ute came to a not-too-sudden rest. Almost like a soft landing.

We sat there in stunned silence for what seemed like an eternity.

I turned off the engine so it would stop screaming.

We looked at each other in complete amazement. It took us a little while to understand what had happened and then we tried to open the doors.

No matter how hard we tried to open the doors to get out they would not budge. They were unlocked but no amount of pushing would force the doors open. It seemed like we were trapped. When the cabin light came on we could see rolls of rusty old wire against the windows.

The worst thing was that no-one knew where we were and being buried in these bundles of wire, we might be in here for some time. I hope not too long as it was my turn to milk the cows in the morning.

I soon learned that the "Old Quarry" where they had taken out thousands of tons of gravel for roadworks and had been filled with rolls upon rolls of old disused fence netting wire. The depth of rolled-up wire softened our fall as we dropped into the mass. It probably saved our lives as a sudden stop would have been like hitting a brick wall at speed but vertically.

The doors were prevented from opening by the wire which had been stored at the bottom of the old quarry. The mate realised at long last that his window was open, so perhaps if we could just work our way through the rolls of wire we might just be okay. So, swinging on the window winder, I opened mine too and we worked at the wire to make our escape.

The old wire was fairly rotten and through the open windows, we could break some of the wire off and bend some more, so we could make a tunnel to climb out of the ute. The mate had found a better way than me so I waited for him to make a big enough hole in the wire large enough to crawl through. As he made his way through the wire, I turned around inside the ute and followed him as he made his way to the surface.

It was tough going. As soon as we put our weight on the wire, it all collapsed and we needed to find a better footing. Some of the wire was sharp too and not only did our jeans get shredded but our legs took a battering as well.

So here we are standing on top of this pile of old rolled up fencing wire looking down at the ute with the headlights still burning. The mate crawled back down again and turned off the lights. No sense in wasting the battery, he reckoned.

We make our way back to his house where we clean up our scratches and tell his parents what had happened. Luckily their house was only about three miles away. It was easy to find in the dark as the back verandah light was left on for him when he got home.

Several days later they took out a crane to remove the ute from the wire. I followed them in our old paddock ute.

You should have seen it. Just the top of the cabin was visible through the wire when we got there. If it hadn't been such a harrowing experience for us we would have spent half the day laughing at the sight of the ute, down there in the hole in the ground buried in old fencing wire. The surrounding wire protruded about two metres above the cabin roof. And the ute was only a few weeks old.

We hitched up the crane and winched the ute out of the hole. It was more than twenty feet down.

Most of the paint had been stripped off as though it had been through a sand-blaster. There was hardly a bent panel but nearly all of the paint was gone. The exhaust pipe was a bit bent but would still do its job. It certainly was not the nice shiny ute that we drove out to the card game the other night.

So off to the panel beaters with it. Just a small amount of panel work, several coats of paint then he was off again a few days later.

So it was decided that from now on we'll just stick to the old paddock ute for the fox work.

Fencing

Ah! Yes! Fencing. I'm glad you asked that question. There are some good yarns about fencing and some of the different ways they can be constructed.

My grandfather used to tell us when we were kids just, how he used to keep the roos out of his cropped area. That was back in the days when he had no money and everything which he did on the farm had to have pretty much no cost or he couldn't do it.

Out of some yet to be cleared bush, he felled some saplings and cut some of the larger ones off at about six feet long, these would be the posts which were stuck into holes set about seven feet apart in the ground to act as posts. The smaller and longer saplings were cut to about fourteen feet and were wound around the posts in a similar way that his dad had told him how hedges were made back in the old country. The tension of the green saplings would hold them tight on the posts. He would make a fence that extended right around his little cropping areas that were little patches of about half an acre amongst the trees. He could not just step over the fences so every so often he would fit a few shorter posts in the ground beside the fence that he could use to tread on so he could step over the fence. Something like an old fashioned style.

He reckoned that if he could maintain a fence height of breast pocket high it would keep out the roos because they would not be able to jump that high. Most of the younger roos and the does were kept out but some of the big bucks were a bit different. A few of them stood at over seven feet tall when they really extended themselves and nothing seemed to stop them. To counter this problem he would camp out near his crops close to harvest and scare off the big buggers with whatever he had at hand. Shouting was okay for a few nights but when laryngitis set in he had to give that idea away. Bashing on the empty tin can worked well if he was close enough to the roos but some of them got used to that method. He tried throwing spears but he just could not seem to perfect the throw that the local abo's had taught him. He tried throwing boomerangs. That was okay but at night he could not see them returning and he got hit on the head several times, so he stopped that too. He could not afford to buy shells for his ten gauge shotgun so he got to refill his old shells with shot and powder, then fit a new cap to each shell. He got quite good at that but when he ran out of shot he would use some pebbles which he collected. Sometimes he just fitted a wad of paper in the shell to hold in the powder, just like a blank.

He only had one accident when he was reloading his own shells. It was hot and in the middle of summer and he had his workshop area under a large gumtree for the shade. He had a shell in its holder ready to load in the shot when a branch fell from the tree. It landed right beside where he was working. It broke the frame of the shell holder and knocked the reloaded shell onto a rock on the ground. The cap must have been a good one because it went off when it hit the rock which was laying there. His leg took some of the blast but he was very upset to have such a large rent in the leg of his favourite work trousers. He did a bit of damage to his hand and in the process lost half his little finger when a shard of rock went flying. A rough bandage saved the rest of the finger and he was back at work on the crosscut saw a few days later. It was several weeks before he got back to the job of reloading his used shells again. The first time back at the bench his hands were shaking but that soon settled as he began working on them.

He would take a kangaroo for his meals every couple of days so that activity with the gun may have scared a few off. But in the summer they come around looking for tucker or water. It would be many years before he had stock but he did have a rainwater tank connected to the roof for his own water or he would cart it from the government bore in town.

As the years progressed he needed better fencing. He lost a couple of horses even though they were hobbled for the night. He searched along the roads and in the bush for days by following their tracks but eventually, he lost those in some rocky ground.

There was a lot of clearing to do and he had invested in a crosscut saw about eight foot long. One man could operate it but it was much more productive with two men. He'd fell a tree and cut it into 6-foot lengths with the saw. Then he would use wedges and a sledgehammer to split the timber into posts about eight inches square. He had lashed out after the last harvest and bought a brace and bit set. The bits ranged from half-inch to four-inch in diameter. Once the posts were cut and split he marked them suitably and drilled four holes each four inches diameter. These were closer to the top of the post and spaced so that when the post was in the ground he would have a hole about a foot from the ground and the rest evenly spaced to the top. Once these posts were rammed tight in their holes and standing, he would measure some smaller round saplings and feed then through the holes, making a very strong post and rail fence. He never lost any more horses after he used that method for his fencing. Where he had to join two rails he would cut the ends at a long angle and slide the ends of two rails into the hole.

It was many years after this that wire became available for farmers to use for fencing. Some of that early fencing wire was terrible. Some were very high tensile and that was great when you needed to get a good strain on a long run, but when it comes time to tie off the end, it would usually break and you'd have to make a long tie.

At least with a plain wire fence, the work required to build it is so much less. The posts can be smaller so they are lighter and you can now carry two at a time. The holes for the wires only need to be about half an inch in diameter. The main difference is that now a strut arrangement is required at each end of the fence so that the wires can be strained up to make it a good fence without pulling over the end post. Some of the early struts were a bit ordinary and were not all that effective. A strainer post with an angled strut post was very common for many plain wire fences, provided that the wire was not strained too tight. In lighter soils and angled strut needed to have a sole plate fitted in the ground to stop the strut post from burying itself into the ground and allowing the strainer post to fall over. Usually, a farmer would make a plain wire fence with six or seven plain wires, all spaced evenly up the post, starting a few inches from the ground. Each wire had to be strained individually starting from the bottom to get the best effect.

There were several ways that a plain wire fence could be strained up. Some cast-iron devices like a cotton reel with holes in it were used. These were put close to the strainer post and the wire passed through a hole near the centre. The holes at each end would allow a pointed tool to be put in and turn the reel to tension the wire. That worked very well. Just drive a four-inch nail through the hole into the post when you've finished and it would stay strained.

About the same time as wire was made available, wire strainers with grips and a chain were used. These work very well too, provided that the operator doesn't swing too hard on the handle.

Some blokes used to tie the horse to the wire with a chain and have him hold the strain while the wire is tied off. That works well too. But it also could cause another problem. If the horse sneezed he could put just too much strain on the wire and cause disastrous effects.

The next major development in fencing was the prefabricated fence. *Ringlock* was one of the first with different manufacturers making their own variations. Basically, these consist of six or seven long wires with vertical wires attached at various spacing and held in place with a little locking ring. Now fencing became an art. The fencing wire was in lengths of a hundred yards. The roll of wire was laid on the ground and you could get the kids to run along after it and unroll the wire beside the posts that you've already put in and bored the holes in them. One end is tied to the strainer post. If that roll didn't reach the other strainer post, another roll of wire was added. To strain this wire you needed a new tool. Two pieces of four by two timber with several bolts to hold the two together with the wire between was very good. After this is fitted to the wire and the bolts tightened with a spanner, the horse would take the strain and the wire tied off. This worked very well

But this could also cause another problem. That of over tension.

The old angle strut was now no good for this fence as they created a fulcrum and many times the strainer would lift out of the ground and the fence would fall down. Not the best outcome when one is trying to control stock. To overcome this problem a new system of strutting at the strainer post was required. This was called the box strut. Now once the strainer post is in place and rammed in tight, a strut post, which is just like an ordinary fence post, is stuck in the ground about six feet from the strainer post in line with the fence. Then a strut rail is fitted between these two very close to the top. Then a brace wire is fitted from the top of the strut post to very low down on the strainer post. This is usually several strands of wire to make it strong and then

tensioned up. This works very well. But it caused another problem.

To tension up this fence, it is best done at hundred-yard sections as these wires are very strong and need to be tensioned evenly. In very later years a tractor or farm ute was used to strain the fence. This works very well but creates another problem. The bloke on the tractor or the ute needed to watch what you were doing while he was driving the vehicle as over tensioning can have disastrous effects. If under tensioned, the fence would be slack and of little value as the stock could just wander straight through the fence. Too much tension may cause the strainer at the other end to come out of the ground.

Have you ever seen a hundred-yard prefabricated fence roll up from the other end and head towards you and over a hundred miles an hour? The bloke in the ute is okay, he has some protection but you are standing out in the open. If you don't get your foot tangled up in the loose wire you just might be able to escape the fast-approaching giant. If you lay down you might get rolled up in the wire too. The only way to escape is to run as fast as you can sideways away from the approaching mess of wire. It has been known to happen on more than one occasion so that is something which one must be mindful of when fencing.

Star steel posts or star pickets as some call them have been around since the 1960s. These can be driven into the ground with a sledgehammer. Later a steel post rammer was made. This consisted of a 2-foot length of 2inch pipe with a plug fitted in one end and handles for gripping on the side. One man can drive in a steel post in a minute or two. Later there has been an engine-driven post driver that works well too. Many contractors have these.

In very recent years some of the more progressing fencing contractors have things very well set up. On a four-wheel-drive truck is mounted a post hole digger at the front. Behind the cab is a stack of posts. Behind this are rolls of wire on end on a braked spindle. It is all operated from the driver's seat. Once the hole is dug, the post is thrown in and a compactor pounds the dirt to tighten the post in the ground. Or a mechanical thumper drives the post into the ground. As the truck drives forward the wire is played out. A bloke at the very back of the truck ties the wire to the posts. A laser and GPS keeps the line neat. If they are using metal posts, then the operation is even simpler.

Wild dogs have become a real problem for many farmers closer to the pastoral regions. Wild dogs are those which have escaped from their owners' homes or left abandoned and have joined up to become packs. They hunt at night and can cause havoc in a flock of sheep or newborn calves. New fences are being installed all the time to alleviate this new menace to farming. Some of these fences are 6 or 7 feet tall.

Electric fencing is another way that farmers have been able to control the movement of stock. This usually works well with cattle, pigs, goats but sometimes sheep can be insulated if they have a heavy fleece. Many city folks have been unwittingly caught by an electric fence. Usually when holidaying at a friend's farm. It was a real chuckle to see their first-ever reaction. They just never seem to fall for the same trick a second time though.

This all works very well, but there is still a problem.

The cost of fencing is now so high that many farmers are not fencing their cropping paddocks.

The First Tractor

Did you ask when we purchased our first tractor? Okay, I'll tell you some background stuff first to lead into my answer.

My grandfather was a great horseman.

His team of Clydesdales were a magnificent bunch of some of the finest horses in the district.

Combine these two together and you would get some idea of how he would operate his farm. Kindness and compassion were his mostly used methods when working with horses. He could get those horses to do almost anything except talk. And sometimes I was told, even that.

Dad did like the horses but he could not work them as well as Granddad did. His temper sometimes got the better of him. The horses worked fairly well for him but by the end of the day, I think they could sense that he was getting tired and would play up. He didn't like shoeing them either. And that should be done each month as the shoes would wear out fairly quickly.

There had been some talk around the place, usually at church on Sundays about a new tractor that was coming on the market soon. Made in England by the Ford Motor Company. These were the Fordson. A twenty horsepower unit with four steel wheels. The two at the front would steer while the two larger wheels at the back would transmit the engine's power to the ground.

Dad had had a good season last year, so off he went to the local machinery dealer to discuss the matter. The year was 1929 and he bought a Fordson Model F. This was imported from England where it was made in the Ford factory there. The dealer was asking four hundred pounds for the unit. After Dad put in the order and paid a deposit of half the total amount he was told that his new tractor should be here in three weeks.

Dad then set about getting a supply of fuel for the tractor. He would need some petrol to get the thing started and kerosene to keep it running. Petrol burns at a higher temperature and is good for starting while the kero produces more power and burns at a cooler temperature allowing the tractor to run for longer periods of time. The local dealer has both of these available for him in five-gallon tins. He brought home several of these so that when he got the tractor he was ready to use it.

The tractor was to replace the large horses for ploughing during seeding, pulling the reaper during harvest and pulling the binder when he was making hay. He reckoned another good virtue of the tractor was that it didn't need to be fed when it was not in use and Dad could spend that time doing other stuff.

I was told that he waited very impatiently till the tractor arrived and was in town half an hour before the dealer opened his door. He had ridden the hack into town and reckoned that it would follow him home when he returned.

But not everything always goes according to plan.

The dealer eventually opened his doors to his shed and welcomed Dad in and showed him the shiny new tractor that he had just bought. It was blue. The larger rear steel wheels had angled cleats riveted to it to give it traction to the ground while the front wheels had a single metal ridge in the centre to allow it to steer more effectively. The seat was made of steel but shaped to suit the shape of most people's backsides, with holes for the seat to breathe. It turned out to be quite comfortable. It was attached to the rest of the tractor with a long flat spring that was supposed to give more comfort to the driver by absorbing the vibrations caused by the steel wheels on the hard ground.

When seated, the driver would straddle the transmission. In the centre of this was a gear stick that he could select any of the four gears. To the left foot are a footplate to rest the foot on and a pedal that was used to operate the clutch. Once pressed down, the gears could be engaged. The right foot also had a footplate and a pedal that applies the brakes to the back wheels when pushed down. Right in front of the driver is a circle of metal with four spokes that join to a central hub which in turn is connected to a shaft that protrudes downwards and forwards. It is at a very comfortable height for the driver to rest his hands on. This, he was told is the steering wheel. When that is rotated to the left the front wheels turn to the left and when rotated to the right the front wheels would turn in that direction. This allows the driver to have a degree of control over which way the front of the tractor would travel and if everything is in order, the rear of the tractor will also follow. The dealer went on, and when you are driving the tractor in reverse turn the steering wheel in the opposite direction.

Right beside the steering wheel is another rod that protrudes. The dealer said that this is the spark advance control. When the lever is pushed down and backwards, the spark of the magneto is retarded. That allows the engine to start more effectively. Once the engine is running and more power is required, move the lever upwards and forwards and the spark is advanced. Now, this is important, the dealer said, never attempt to crank the engine when the spark is advanced or it will certainly backfire on you and that can be a very unpleasant experience.

To the right of the steering wheel is another rod that is similar to the spark control. That is the throttle, said the dealer. Move that towards you and downwards and the engine will speed up. Move it away and the engine will slow down. The little lever beneath the throttle is the choke control. You will need to pull this out when starting as it allows more petrol to be fed into the engine at start-up when swinging on the crank handle. But as the engine warms up it can be pushed in to lean off the mixture. He showed Dad where all of these controls extend to and where he can adjust them from the ground when starting the engine.

The one electric switch that is below the steering wheel is the ignition switch. To allow the engine to start that needs to be turned on. Move the switch lever downwards for on and move it upwards to stop the engine.

That was just so complicated and it took the dealer several times to get the information across to Dad. But after several hours the dealer was satisfied that Dad had enough grasp of the situation to be able to drive the thing home and have some time there to get used to it.

So he did.

He drove home.

But first, he had to start the engine. The dealer explained that it must always start on petrol or it just would not start. At the front of the engine is an angled piece of metal rod protruding from low down. This turned out to be the crank handle. Now, Dad had heard of a few cranks in his time but never one just like that. The dealer showed him how to hold the handle with all fingers on the leading side so that in the event of the engine backfiring the crank handle would swing away from his hand and not hurt him. He did say that there are several blokes around who didn't do this and have broken thumbs that prevented them from working for a few weeks.

Dad set the ignition switch, the spark, choke and throttle just as the dealer had shown him and after grasping the crank handle, swung it around. The engine fired on the first revolution, much to the delight of Dad.

That was good but it created several problems. It was very noisy. The dealer suggested that Dad should drive it out of the shed, so after jumping onto the seat, he selected a low forward gear. He released the clutch and the tractor lurched forward. The dealer told him that it will take a little practice to have a smooth departure. As he drove out of the building the noise abated, much to his delight. He steered the tractor to the road that would take him home and headed off down that road, advancing the throttle as he went, causing the tractor to speed up to its maximum of ten miles per hour.

After a few minutes, he stopped.

The horse!

Where is the horse? He had ridden it to the dealers and had left it tied to the hitching rail that was beside the main building.

It was not there.

A bloke who was standing across the road ran down the road to tell Dad that as soon as the tractor had started, it reared up and took off in fright down the road heading in the direction of home with its tail flying in the air.

So the noise of the engine had startled the horse! That would be something to consider for the future. Dad continued on home without further incident and found the horse beside the stables. He did stop short so that it wouldn't get spooked again and he put the horse away for the night after a good brush down.

He parked the tractor near the house remembering to turn the kero tap off and the petrol tap on for a few minutes to allow the petrol to fill the carburettor bowl, ready for a start in the morning.

So far so good.

That was Dad's first introduction into things mechanical. After a few days driving the tractor around doing different jobs he soon got the hang of the controls and became a very efficient driver.

The next seasons saw him working his farm with the tractor instead of the horses which had been used for the last half a century.

One rather wet year saw most of the tractors in the district become bogged as the soil would not take the weight of them with such a small footprint. Different blokes had worked out many different ways to un-bog tractors. Some even pulled their tractors out with a team of horses but this was frowned on as "not the thing to do." Some put bushes, sticks or logs under the wheels to give them more floatation and this worked too.

Dad came up with a brilliant idea. If floatation is the issue, why not increase the floatation of the tractor by bolting some cut-off sleepers across the rear wheels. After thinking on this problem for several days, he cut some timbers to about three feet in length. These he bored holes in them near the centres and fitted them right around the rear wheels and parallel to the axle. It took twenty-two pieces to go right around each wheel. So now instead of the wheel being just one foot wide, it is three feet wide. That should stop the tractor from getting bogged any more. The next three weeks of seeding went very well and he was able to get to those wet spots much easier than before, without the tractor going down.

But one day the ground opened up and the tractor went down. It sat neatly on the diff. Dad immediately pushed in the clutch to stop it from digging in any further. After sitting there and pondering the situation further he reckoned that if he now had so much traction he should be able to just drive the tractor out of the bog.

NOT SO.

After disconnecting the plough, he carefully selected bottom gear and eased out the clutch. For a few seconds, nothing happened as the clutch continued to slip. Gradually the clutch gained a grip and instead of the tractor moving forwards, the front of the tractor reared up.

He immediately declutched and threw the gear into neutral. The front of the tractor came crashing down again. The rear wheels had become so stuck in the mud that the tractor tried to turn itself around the rear axle instead of the other way around. Had the tractor continued on its path, the front of the tractor would have rotated right over and landed upside down on top of Dad as he sat in the seat, bringing him to an early demise.

His quick thinking had saved his life.

Several other people had bad experiences with their new tractors too, particularly when working on the steeper of slopes. Tractor roll-overs have maimed or killed many operators over the years until the tractor manufacturers made them a little more stable.

But that was Dad's first tractor and he has had many since that time. Each was traded in to the dealer as a new one was purchased. Apart from not having so much old machinery around the yard he also saved some on the cost of the new tractor.

The First Truck

Dad told of the times when he and granddad drove their team and wagon to take a load of bagged grain to the wharf, down by the coast. The cost of sending the harvest by rail had increased and Granddad was determined to see if he could deliver his load of bagged grain from the farm to the port for export at a better price than either the cartage contractor or the railways could deliver it. Granddad tried with a load on his old wagon that shows a lot of signs of wear.

They had help from a neighbour to load the wagon with two hundred and twenty bags of wheat. The weight of the load was about eleven tons. Before loading the wheat, Granddad had jacked up each wheel, removed it from the axle and applied new grease.

The load was going to be heavy.

The road to the coast was hilly and with many bends. Dad was able to ride with him as a ten-year-old lad and he remembers that trip oh so very well. It took them four days to cover the fifteen miles.

The coast was three hundred feet lower than their farm, so mostly they were going downhill. But as with so many roads in that hilly district, they all seem to go uphill too and that is what made this trip so memorable.

They started out just before dawn on a Monday. The horses were well-rested, well-fed and were behaving well, so they thought that they would have such a breeze of a trip to the coast with this large load. Granddad was not all that used to driving a team of eight Clydesdales with such a heavy load and had listened to what some of the older drivers had told him but Granddad didn't tell Dad this until after the event. Granddad even had six spare horses tied to the wagon with the harness for them too. The reason for this became evident well into the trip.

Going along the flat ground was easy. Granddad took his seat on top of the bags of grain and Dad walked alongside, checking to see that everything was going well with the wagon. Granddad had all of the reins in his grasp so that he could control the horses and if it was necessary Dad would run-up to the head of the lead horses and quiet them down.

Last week Granddad had the team working together several times and he was happy that he had the right horses in the lead and the right horses in the shafts.

Several of the horses in between were a little bit skittish but after several hours of training they settled down into an easy fast walk.

Now that the load was on them they seemed to settle into the pace very well and Granddad even had a smile on his face as they travelled along.

The ground began to slope upwards and the horses began to strain into their harness as they pulled with all their might to keep that wagon rolling. It wasn't a steep slope but "enough to make them earn their keep," Granddad said.

At the top of the rise, Granddad called for a halt. The horses were glad to stop as several of them were blowing a bit harder than usual.

Granddad pointed out that the track from here for a few miles was all downhill and he was quite sure that the hand brakes on the old wagon would not hold the load should they need to stop or slow down. They carried a full set of chocks for each of the large steel tyred wooden wheels of the wagon so that they could hold it steady on a slope. Granddad instructed Dad very carefully on how to fit these in front of and behind the wheels to prevent them from moving. At this stage, Granddad climbed down from his high perch on top of the load and the two of them untied the four lead horses. One started dancing around but they soon had the harness and tackle untangled. These horses were then attached to the rear of the wagon and hitched to the brake attachment points fitted to it with their trace chains. Granddad explained to Dad that these four horses should be enough to activate the brakes on the wheels to slow the wagon down enough to keep it steady and allow the horses in the shafts to be able to steer the wagon. The six spare horses were added to those which he had taken from the front and attached them with trace chains too. Now they had ten horses as the brake horses.

The brake attachment point is a heavy lever with a fulcrum point close to the wheel and it had another piece of timber attached to it that was shaped the same as the wheel. It was, in fact, a brake shoe. When the horses pulled this from the back of the wagon the strain on the attachment point would cause the brake shoes to engage on the wheel. The more pressure on the brake shoes the more stopping power they had. Unless the wheel should lockup and skid.

A good theory.

He encouraged the ten brake horses to take the strain backwards. When Granddad released the brakes he had Dad remove the chocks from behind the wheels and throw them into the bin under the belly of the wagon. With the tension that the rear horses applied, the wagon stayed still on the gentle slope.

Now everything looked good for the descent.

By carefully balancing the forward movement of the brake horses, the wagon would begin to move forward. Granddad had to make sure that the shafters would not think they had an easy job either so he kept just enough pressure on the brakes so that the shafters would have some work to do. That should keep them all happy. It took Granddad and Dad some time to get that sorted and then keep the balance correct. A variation to the slope required some trial and error but they got it all working smoothly.

Then they came to the bottom of the slope.

Once again they stopped the wagon and Dad fitted the chocks to the huge wheels. The four horses which used to be in the lead are now untied from the rear of the wagon and put back in front again. They would be needed to help the shafters pull the wagon up this slope.

They got going again and all was going well. The spare horses just followed along as though they were on holiday, but that was to come to an end shortly.

Just around the next corner, the slope became a hill. The spare horses are now required to be put at the front. The lead horses were untied from their position and the spares chained in, then the lead horses attached to the front once again. Now all of the horses are in front of the wagon and shortly were pulling the heavy load up the hill.

Now, Dad only had to run from the wagon forward to attend to any horse that was misbehaving.

This process was completed over forty times on that trip to the coast. That is twenty hills and twenty dales. They did have a few problems but these were only minor and quickly overcome by these resourceful blokes. Once the wagon was unloaded, the return trip began. The same process was followed but the spare horses were swapped with some of the others to give them a rest.

With this in his mind so many years later, Dad decided that he needed to keep up with the modernisation of the farm. The old wagon was now broken down after it had been modified to be pulled by a tractor but it was just so heavy. A smaller wagon was more suited to the tractor as it drove along quite a bit faster than the team of horses.

The machinery dealer in town had told Dad of these new Ford trucks that were available and after Dad had another good cropping season he decided to pick up a truck for the farm. He had thought of a used one, but they were all worn out and would need repairs, so he decided that he would get a shiny new truck. His mechanical knowledge had improved so much after he bought the tractor so now he was prepared for what was to come.

The same dealer who sold him his first tractor had a shiny new green and black truck in his shop. It was made by Ford and imported from America. It would carry 8 tons and it  was powered by a side-valve engine with eight cylinders arranged in a "Vee" formation. It still needed a crank handle to start, but Dad was quite adept at doing that now. He did strain his thumb one day when he was thinking about something else when cranking the engine and it backfired on him. Luckily he moved his hand out of the way very quickly.

The bank came to the party with some loaned money and soon Dad had his new truck. It was just so much better for carting the bags of grain to the siding and it didn't need to be fed when it wasn't in use, either. It usually stood still when you wanted it to and didn't wander off like one of the horses used to.

Fuel was now available in the larger forty-four-gallon drums and Dad would bring home some of these when he took a load of grain into the railway yards for transport to the markets.

Now that the farm has some mechanisation it became evident that he and other farmers did not need to employ so many people at harvest time and this caused several families to leave the district in search for other employment. Dad was sorry that this should happen but he needed to make a profit from his enterprise or the bank would be after him and he had worked too hard for too many years for him to allow that to happen.

As well as cropping Dad was also running a few hundred head of cattle and had a good breeding program in progress so with a stock crate fitted to the back of the truck he could now transport his own cattle to the markets too.

Yes, the old Ford truck did a marvellous job for many years and I think he has had four or five trucks like that since.

Dad used to relate the story of Granddad's cousins who, in the early years, had several teams of bullocks. The boys were very hard workers and had a steam engine and a chaff cutter. They would travel around the surrounding districts and cut the harvested sheaves and straw into chaff for the farmers.

They worked for so many different farmers that they lost count of how many farms they worked on over that period of time but they were always welcomed back for the next harvest.

Now, when you hear mentioned that a bullocky is coming down the road with his team it was usual for the men to take their womenfolk and children away from them as the bullocky was known to use the foulest of language when he was handling his bullocks. He seemed to think that by swearing at them he could get them to work harder. Or he was not so good at controlling his temper.

Granddad's cousins were also Christian folk and did not use the foul language that other bullockies were renowned for. They cared for their animals and treated them with respect. This had proved several times that they were able to get more from their teams than most of the other bullockies were able.

The steam engine and chaff cutter were shifted around on their own wheels and they had a wagon with all of the necessary items that they needed to do the job. So the three teams were about half a mile long when they were all travelling along the roads at the same time. When negotiating a hill, one wagon was taken at a time with the bullocks from the other wagons in harness assisting. Eventually, the three units would get to the top of the hill and then the same procedure was followed to travel on the downward run but the bullocks were chained to the back to act as brakes.

On some of the unmade roads, this wasn't necessary on the downward run as the mud was a good enough brake except on the steepest hills. But the uphill run was much heavier.

When they had the steam engine set up, they would work with the farmer to cart the straw to the chaff cutter and also to cart fire-wood and water in order to keep the boiler well fed to maintain a good head of steam.

It was very hard work, feeding the boiler, forking in the straw, fitting the empty bags onto the bagging chute and then lifting away the filled bags and stacking them for the farmer. Sometimes they would make a stack in the paddock and sometimes they would stack the full bags onto a wagon for transport.

On some farms, they would cut 40 tons of chaff but for some of the larger farmers they cut over 200 tons of chaff in one season. There was no mention of the total annual tonnage of chaff which they cut but with some good growing seasons, it must have amounted to many thousands of tons.

By the time that mechanised transport was available, they had progressed to other means of income.

Going Bulk

When did we stop using bags for our fertiliser and grain? Well, now that's a bit of a yarn, so I hope that you'll bear with me for a while so I can give you some background detail.

What we've discussed in the past chapters may have given you some insight into the handling of our produce in bags, let me give you some of the finer details.

When great-grandfather began farming, the old jute bag was all that was available for the storing and handling of the grain that was harvested.

His first few years were difficult as he could not afford to buy too many bags at one time. They were just so expensive at threepence each. Doesn't sound much now but it was the price of a loaf of bread if you bought it from the local baker.

So what did he do?

He dug a hole in the floor of his bough shed and lined this with some old hessian that he had used for another purpose and after filling all of the bags that he had, he would shovel the remaining grain into the hole.

Those filled bags he loaded onto his old wagon or had the neighbour take them for him.

When his grain was delivered to the siding, the company which bought it from him would allocate him another lot of bags, so he took these back to the farm and filled these from the hole in the floor.

The threshed grain was spread out on the floor and had to be shovelled by hand into the bags or the pit as more grain was threshed by hand from the remaining stalks.

Shovelling the grain into the pit was a fairly simple and easy job but standing in the bottom of the pit to shovel the grain out again was a bit of a task, particularly if you wanted the grain to make its way into the bag at the same time.

Great-Grandfather was a bit of an inventor so he arranged a sheet of iron on a slope above the pit and shovelled the grain from the pit onto this sheet of iron that formed a funnel and most of the grain found its way into the bag.

This worked quite well but created another problem.

It was so far up in the air to get enough slope to get the grain into the bag, it was hard work. So he dug another hole for the bag to sit in.

Now he didn't have so far to throw the grain from the shovel.

Needless to say that by the end of the long working day, he was quite tired.

After several weeks of working this way, he had another idea.

He set up a windlass above the pit and took a bag down with him to fill. This he did with ease. Then he attached the rope to the bag and after he jumped out of the pit would wind the windlass handle and raise the filled bag to the surface.

This worked well.

To make this job a bit easier he collected a few light bush poles and made a ladder to climb in and out of the pit.

But there were a few pitfalls with this system. The bough shed with the grain-pit was on a bit of a slope. One night just about midnight there was a violent thunderstorm and the rain was so heavy and coming sideways that Great-granddad could hardly walk straight as he raced headlong to the shed. He knew that water would run down the slope and into the pit if he didn't get there fast enough to dig a drain to take the water away. He knows that he should have done this job as soon as he made the pit but other jobs kept getting in the way. He began feverishly digging a drain but was only halfway done when the water began running from the slope down into the shed. The rain just did not let up and in a few minutes, he was drenched right through to the skin.

He kept on digging and shovelling the soil but if there were about ten of him he might have stood a chance but just by himself the job just could not be completed in time. The water had run into the pit and after just a short time the pit was full to overflowing. The grain which was in the pit, being buoyant, began to flow out the other side of the shed.

Now he had another problem.

If the horses began feeding on that grain, they would become bloated and die. And that would take just an hour or so.

He now had to shift the horses into a yard that was uphill from the shed. Luckily they were all smart enough to shelter out of the worst of the rain in the little bough shed alongside the one with the pit. He closed the gate to contain them. Presently he found the bridles after scrounging around in the dark. He fitted these to the horses and two at a time he walked them to the uphill yard. The horses were none too happy at being shifted in such a manner and complained loudly but he was able to get them all shifted just as the sun was rising.

After day-break, he stood back and surveyed his night-time efforts and was astounded at just how far the grain had spread.

At least he had finished harvesting yesterday or what grain was still standing would be ruined anyway. Now he just has a massive clean-up job to do.

For the most part, this system worked well for him for many years until he was able to purchase his threshing machine. Now the bags stood under the outlet of the threshing machine and the filling of the bags is so much easier.

But it was still a very manual system.

Next, he bought that new Ridley stripper which took the grain from the standing crop and the bags were filled from under the outlet of the hopper while the machine was moving along.

This worked well for Great-Grandfather but created several other problems.

Now he needed more bags.

And he needed two men to stand at the bagging outlet to change the bags when they were full and stitch the tops of them, while he drove his team of horses on the front of the new machine.

All of a sudden, these new methods of farming had become more expensive.

And now he had to crop a larger area to cover his costs.

That meant he had to purchase more land and develop it.

Great-Grandfather loved his horses and it took him some time before he could bring himself to

replace them with the more modern machinery which was becoming available for farmers like him to use on their properties.
He did give in but not until long after Granddad joined him.

They still continued on in this manner even after my Granddad joined him in the late 1920s when he left school at the age of 12.
Soon he had that new tractor that we talked about and he attached the stripper to the rear of it. That worked well for a few years before it was replaced with a Sunshine harvester. Now they could cover so much more ground when he was harvesting. But he still needed two men at the bagging outlet.

My Dad worked with Granddad after Great-granddad passed away, for many years in this same manner. After years, Dad could see that there must be a better way of handling the grain at harvest time and he spent some time playing around with the new idea of using augers for lifting grain.
Soon they had a larger bin or hopper on the header *(this is what the new Sunshine harvester is*

called) to carry a larger amount of grain. After an hour of harvesting, he would drive the header over to a tank with a conical bottom and auger the grain from the header's bin into it. The auger was fitted with a small petrol engine to operate it.

The tank with the conical bottom was referred to as a field bin and once filled, the bags would be fitted to its outlet from where they would be filled. Now that this job is done stationary it was so much easier to do.

It was about this time that I joined them on the farm and we had the three generations working together again. I had seen some improvements to agriculture while I was at an agricultural school.

But the bags are still heavy and they still had to be loaded by hand onto the truck to be delivered to the siding where they would be unloaded onto the waiting rail wagon.

Still a lot of manual handling.

The truck was nearly worn out with so many trips carting the grain to the railhead and soon a brand new Dodge was bought to carry up to 8 tons of grain.

It didn't take long before the idea of bulk carting bins was bandied about by the local farmer's group.

After some encouragement, the grain purchasers took on the idea of bulk handing.

Now Dad and I fitted a bin to the back of the truck.

This bin had a "vee" shaped bottom with an auger running across the vee to deliver the grain out the side of the bin.

Now the truck could follow the header around the paddock and take the grain from the header's auger straight in with no handling at all. Then it was just a simple job of driving to the rail wagon and using this new system, auger the grain straight into the wagon.

One advantage with this new bulk handling system on the back of the truck is that now we could cart all of the fertiliser that was needed for the cropping program as well. This fertiliser was unloaded straight into the seeder or into a covered shed to be used later on.

No more bags.

The system improved over the years and now they don't run Dodge trucks with small side-emptying bins any more.
They have these massive 700 horsepower trucks with two 40 foot tipping bins on trailers behind.
The old header is now replaced by a massive thing that cuts a 50-foot swath of crop at a time and handles as much grain in an hour as grandad used to harvest in a year.
The cost, you ask?
Well yes, the cost is astronomical!
Great-Grandads shovel cost him 2 shillings when he started and now the harvesting system costs around a million dollars.
It's any wonder the banks are so rich.

Fire

Yes, I can remember being involved in fighting several fires. I've been part of the local volunteer firefighting group for most of my life but I did retire from active service a couple of years ago but still get to the events and meetings.

But back in my grandad's day, they didn't have all the fancy gear that is currently available to our firefighters.

When I was a lad we used to help Granddad with wheat bags and the covers from the woolpacks to fight fires. It worked quite well on a paddock fire. Granddad would drive the horse and dray with a small tank of water on the back. We'd throw the dried bags up to him and he would keep the bags wet from the open tank on the back of the dray. He would throw them down to us, who mostly would catch them. Sometimes he would surprise us by pointing to some activity close by then carefully drop the wet bag on our heads. At least he thought it was a great laugh.

We'd take the wet bags and swing them at the flames to extinguish them.

Sometimes when the tank runs out of water, we'd use shovels and rakes to make a firebreak and this was alright in a paddock fire but not of much use in a fire in the bush.

The main secret here was to get to the fire as soon as it starts. It was so much easier to get the fire under control while it was still small. Once it got bigger it was so much more difficult to control.

At one stage, a Furphy tank on wheels with shafts was harnessed to the horse and this had a hand pump on it. Dad or Granddad would drive the horse and if the horse wasn't too frightened by the fire, he would stand while Granddad swung on the pump handle. A long hose ran out from the pump and one of us would direct the stream of water at the fire.

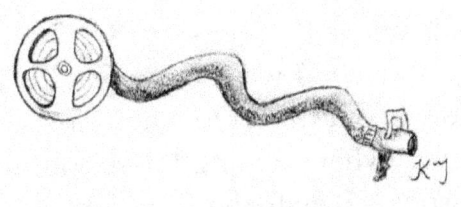

Later on, grandma would drive the old farm ute with a tank on the back, filled with water. This was also fitted with a hand pump and it worked well. We used to fight over who should turn the handle on the pump, as that was so much easier than dragging that heavy hose around but either

Dad or Granddad ended up at the pump, riding along on the back of the ute.

In later years, in some of the hilly country, we'd use one of the small tractors with a tank on the carry-all and this worked well too. The tractor has a power take-off and to this was fitted a small gear pump and when it's connected to the tank and a longer hose connected to the nozzle, that works well too.

On some of the flat country, several of our neighbours have old ex-army four-wheel-drive trucks which were fitted with tanks, pumps and long hoses and these worked well too.

Until they become bogged.

I saw one situation when I was quite young, we were at a fire and had our tractor towing a fire cart and could get everywhere, diving in and out of some of the hot-spots to get to the seat of the fires but the trucks could get a lot done in the more open country. This particular day the paddock fire, which I think was started from a wheat stripper that had nearly finished the paddock, got away into the bush beside the dry paddock. It hadn't rained for a few months and the bush was very dry. Once the fire got into that bush, the smoke became very dark and if you were in the wrong place visibility and breathing became very difficult.

The previous year had been very dry and this was the first good crop that we had had for several years and some dry sand had blown up against the edge of the bushes and along the fence-line. The driver of the truck was not aware of the drift-sand and tried to drive along the fence-line the get ahead of the flames so that they could apply the squeeze motion. It nearly got there too. About fifty yards from the seat of the fire the truck got stuck in the soft sand. Three of the blokes on the back of the truck jumped off to let some air out of the tyres to improve the floatation and get the truck out of the boggy sand but before they could get their job done, the wind changed and the fire turned back on them.

It was about the time when John Landy performed his miracle mile at the Olympics but I reckon that if he was here, he would have come second.

Boy those blokes could run! In full kit too!

With their previous experience, they had the forethought to run across the wind and were able to escape the fire but they lost the truck. We watched as it burned with loud bangs as each tyre exploded and then a big one as the petrol tanks went up.

It looked like cracker night all over again.

We never lost a man in our district while fighting fires but some other districts were not so lucky. I think much of the reason for this is that our families had been dealing with these sorts of fires for many years and learnt to respect what a fire can do and that you never know when the wind will change and force the fire back onto the firefighter.

That knowledge only comes with experience.

Every summer we would attend three or four fires in our local district and our local council had the foresight to enforce all landowners in the district to install firebreaks around their properties.

A disc plough or scarifier behind the tractor was very good

for this but in later years some farmers resorted to the use of chemicals to keep the grasses from growing at their boundaries.

We were required to clean all combustible material for ten feet away from any buildings and we even used to burn off along the roadside of our busy road. This is what kept our district almost fire-free. We even ploughed a fire break around our haystacks to keep them safe and that was effective on more than one occasion.

But then, there's always that idiot who flicks his burning cigarette butt out the window of his car as he drives along without any thought of what catastrophe it may cause.

Some people just don't think or are stupid, I am still not sure which.

There was a very big fire in our adjoining district and it took several weeks to get it under control. The fire burnt out two townships and many farms lost everything. It seems that the fire started from a lightning strike up in the big bush area that was quite inaccessible and it grew ferocious very quickly due to the dry weather and strong winds. There were all sorts of equipment used to control that fire-beast. I remember Dad was out with our fire unit for much of the time and we would only see him for a few minutes every few days. The rest of us had to keep the farm going and milking the cows while he was away, Mum had someone look after the house and she helped us too. Our sister was even called back from her nurse training to help.

It was all hands on deck.

At about the time that I was retiring from the farm, the government came up with the idea that they would make a government department to take control of the firefighters. We were all volunteers and could see the stupidity of what they were doing. But after about ten years they began to see the problem from our point of view and come to some sort of common-sense arrangement, making a few changes.

There is nothing funny about fires as they can be so final and so devastating but I'll always remember those poor blokes running towards us as their fire truck burned to the ground. A few weeks after that fire event, we helped them build up another truck and while we had a break one afternoon, we told them about how we saw the event unfolding. Yes, it was devastating and emotional but with a bit of rousing, they began to see the funny side as we had seen their escape and we all had a good laugh about it.

Milking Cows

I'm glad you asked about dairy cows. Like most farmers, Granddad and Grandma had a dairy cow too.

It was a cranky old thing.

Every time Granddad went to milk her, she would raise her leg and kick at him, sometimes even making contact if he was a little too slow to move out of the way.

Granddad was a pleasant old soul and usually put up with those little discomforts that happen from time to time.

The old cow would be held in a paddock close to the house so he didn't have too far to send Grandma to fetch her for milking and sometimes he would milk the old cow himself.

The old cow never seemed to mind when Grandma milked her and she stood ever so still for her. Never moving a muscle or even swishing her tail at the odd fly that should wander too close.

Grandma's bucket was always fuller than his was when he milked her. The old cow would just stand there chewing her cud after gathering up the remains of the crushed oats that she was fed in the feeder.

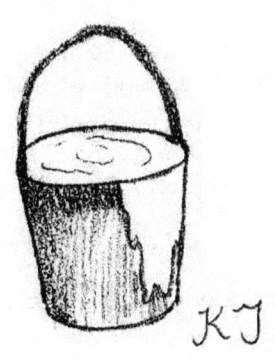

Even when Granddad milked the old cow, usually grandma would have to put her in the bail with the dipper full of feed, as she just would not behave when Granddad tried to put her in.

It was always a bit of a scene. Twice a day they would repeat the same process, Grandma bringing in the old cow and then sometimes it was Grandad sitting on the old three-legged stool, hand milking her.

This one day that is of note, a thunderstorm was approaching and the smell of rain was in the air. The thunder was rolling high up in the clouds many miles away and would seemingly pose no threat to them. But the old cow seemed to have a different view of the matter.

Granddad had just sat down when she lifted her leg in the usual fashion. He braced himself against the movement of the leg and all the while never missing a beat with the pulling and squeezing of the teats. The milk flowed in spurts into the bucket.

That was until there was a louder clap of thunder that seemed to be just across the paddock.

The old cow jumped.

Granddad jumped.

The cow lifted her leg.

The cow's foot caught the edge of the bucket and she began swishing her tail even though there were no flies around. The bucket which was nearly as full as it had ever been, spilled.

Granddad hung on tight to the bucket to save as much milk as possible while still grasping one of her teats. The way he held his balance you'd have thought that he would have a future in the circus. The cow's long tail had been dragging in some fresh, wet manure and the tuft at the end was very wet, smelly and dirty. A swish of the tail landed squarely full in Granddad's face with as much force as the old cow could muster.

It must have been about then when he lost his temper.

He lost his balance and fell off his stool, landing flat on his back, sprouting some words that Grandma had never heard before. He must have learned them all those years ago when he spent some time working with the old "bullocky".

The bucket that he was hanging onto came off the ground and emptied itself all over him leaving the bucket covering his head. It may have had a little help from the old cow's swinging hoof.

Granddad didn't lay there for very long. He scrambled to his feet with a continuous stream of insults for the old cow, shaking most of the spilled milk from his clothes. The old bullocky words just seemed to come streaming out, seemingly without any effort by him.

After a time his reddened face softened and he began to relax again and return to his old self.

In the end, he very quietly walked up to the bail that had held the cow's head restrained and gently released her from its grasp. The old cow let out a loud bellow as if to scold him for what had happened.

Granddad never milked that cranky old cow again after that event and Grandma had to do the entire job by herself.

"Anyway", he told Grandma, "there are so many other things for me to do".

Granddad's cousin Charlie was a dairy farmer who milked 30 to 40 cows by hand twice a day every day, seven days a week. Some of these cows were yellow, some were red, and some were black and white. It was quite a sight to behold, watching them walk in a line towards the dairy at milking time.

Charlie sometimes had the help of his wife, Margaret but she became busy bringing up their 4 boys and could not seem to take the time to be there for him all the time.

As the boys got a bit older, I would visit them from time to time. Due to Dad's farming practices, this would be after the crop was planted and before it was harvested. So basically while the crop was growing, he didn't seem to need me quite so much then.

At this time of the year, it was usually cold. As I was about the same age as Uncle Charlie's eldest lad Bill, we spent some time together.

We were always adventuresome and looking for things to do, so Uncle Charlie had Bill and I bring in the cows from the pasture paddock for the milking. It was great in the afternoon as we could see in the daylight and sort of getting a bit sidetracked on the way. There was just so much for a young mind to take in back then.

Things like firing the stone from the ging at the crows to get them to move on. They were always a problem pinching the nearly ripe fruit from the trees in the orchard.

Or chasing a rabbit down its hole. There were a lot of them around then but the ten-eighty poison became very effective later on.

Or collecting an egg or two from the magpie's nest. They'd swoop us from time to time but we always wore our hats so they didn't take off too much skin in their attacks.

The mornings were the worst.

We had to get up by about 4:30. It was dark and usually bitterly cold when that breeze blew in from the south. It must have started down near Antarctica somewhere, it was so cold. And we didn't wear boots because they would get soaked on the wet grass anyway.

Our feet froze.

But after a time we learned to seek out the fresh cow dung, particularly if you didn't chase the cows too hard, they would leave their dropping in a nice little cow-pat.

And it was warm.

We would find a nice fresh one and spend a minute or so warming our feet before moving on to find the next little steaming pile. Aunty always made us wash our feet before coming in for breakfast, so she must have known what we were doing.

After some years passed, Uncle Charlie found that to keep up with the rising costs of his farm, it was necessary to increase the size of his herd. He did this over several years but it was just after a small accident that he had to employ a man or sometimes a girl to work in the dairy for him. This, of course, increased his costs.

He faced the old dilemma that nearly every farmer faces. Does he get bigger to cover costs to make a reasonable income, or is it better to let the cows go and move into some other branch of farming? He loved his cows and he loved the land, so moving from the farm was not an option.

I remember him talking this over with Dad a few times.

The next thing we see is Uncle Charlie having a new dairy built. It was a fancy one too. With this design, two or three people could milk 250 cows in 2 hours by themselves.

At the end of that year, the dairy was completed alongside the old dairy which was kept operating at the same time. There was no more hand milking from then on as the new dairy was complete with a diesel-powered vacuum milking machine. Very fancy it was and one of the first in the district too.

In the past when the cows were milked and the milk filled the ten-gallon cans, lids were fitted and the cans were taken down to the creek to lay in the cool water. This cooled the milk and just about morning tea time, they would pull the cans from the creek and load them onto the old Ford model A truck and drive down to the front gate and offload them onto the ramp where the cartage contractor would collect them and deliver them to the dairy processor by the city. A couple of old chaff bags soaked in the creek water was thrown over the top to keep them cool until he arrived.

Nowadays the milk is run into a 500-gallon refrigerated vat and is collected straight from the dairy by a bulk road tanker.

Boy! How things have changed.

A neighbour of his had passed away and the family wanted to move on and offered their farm to Uncle Charlie at a good price.

Uncle Charlie spent some time with the local bank manager and soon he had enough of a loan to increase his acres and buy some more dairy cows and a good bull for his breeding program.

All he had to do now was get his dairy into full operation so that he could pay back the bank.

He would often comment when we visited, that he seemed to think that the bank was making more money out of his dairy farm then he was.

But Uncle Charlie and his two boys kept on with the dairy and it eventually turned out to be very successful.

But to make the best of what they had, they had to diversify some. They began making and selling cheese and home-made butter.

At every "farmers market" around the district, the boys would be there with their butter and cheese. People just seemed to flock to them on their stand, with the refrigerated trailer standing behind. I think those people could detect the quality of their produce.

I'm really not sure if the tax-man knew just how much cheese and butter they sold, as those markets are mostly cash sales.

But good luck to them, I say!

I heard the other day that Uncle Charlie's boys had done very well with their dairy. Of course, they are about the same age as me and retired now. Their sons didn't want to do the long hours and hard work of the farm, so they have moved on but Don's daughter has been very keen on the dairy ever since she was a nipper and now she runs the outfit with her husband.

I hear she is doing very well too! She has a few clues indeed.

Once again with the bank's help, they had just built a new dairy on the farm, not far from where the one that Uncle Charlie had built about 50 years ago.

Now, this is a fancy dairy.

All semi-automatic, all-electric system with a rotating section in the middle holding about 45 cows at a time. The cows would come in and put their heads into the feed bin. The back end of the cows then faced a pit where the operator stands and every minute or so he puts the teat cups on the next cow's udder and milking begins. The platform rotates and by the time the cow has let down all of her milk, the machine removes the cups all by itself. The platform continues going around just a bit more and the cow walks out of the shed and off to her day paddock. The same process is repeated at the end of the day, so the cows are still milked twice a day.

Now they are milking around 750 cows and I'm told that as this is one of the most modern dairies in the country. It can milk up to 1200 cows each time, so there is room for expansion in the future.

The Pigs

You asked if there were pigs involved in our family's history!

Well, yes, of course, there were.

Way back in time when Granddad and Grandma were milking that grumpy old cow, they had a few pigs.

Grandma would save the milk in the old Coolgardie safe. It kept well there for three or four days in the wintertime but in the summertime, they were lucky to have it last for two days before it went off.

While the milk was in the old safe trying to keep cool, the cream would rise to the top and she would remove this and after a few days of milking would have enough cream to make some butter.

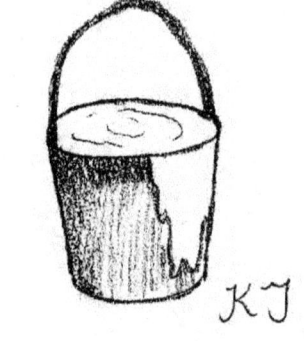

An old butter churn was used here and Grandma would put in the cream and add a bit of salt to the mixture and begin turning the handle. After a time the mixture would thicken and with a pair of "butter-pats", she would form the lump of yellow mass into a block of butter and wrap it in grease-proof paper. After this cured in the cooler for a day it was ready to use.

She was a wiz at making bread too and when a slice of her bread was slathered with a slab of her butter, it was something to write home about. The butter tastes nothing like that manufactured stuff that we get in the supermarkets today. It certainly had its own very rich creamy texture that tasted wonderful. When we visited, we always asked for more.

The milk was used in their porridge at breakfast time and tea or coffee during the day and in their custard at mealtime and in their hot chocolate at night. What was leftover at the end of the day was fed to the pigs.

The pigs also had a supply of kitchen scraps but sometimes there wasn't much of these either. Granddad had some grain that he would soak in water for a few days before feeding this gruel to the pigs.

Grandad had built a sty for the pigs out of some heavy timber that kept them under control. There were two yards and a shelter that kept them out of the sun in the summer and the rain in the winter. The yards led to a small paddock that was well fenced and there was a wallow for the pigs at the bottom of the paddock.

It was amazing to see those big sows wallowing in the mud when the temperature was around 1 or 2 degrees in the winter mornings. In the summertime, of course, they needed the mud and water that was in the wallow to keep cool as we all know that pigs don't sweat, so that is one way for them to keep cool in the heat. Granddad set up a tank to store some water to keep the wallow moist all the time. He has several drums on the back of the dray that he used to cart the water. Usually, the windmill would give them enough water but there were times that he had to cart the water from the government bore and tank about 5 miles down the road towards town.

The tank which was connected to the windmill was on a stand so he just drove the horses to place the drums under the outlet and he'd fill the drums from the hose-pipe.

But when he needed to collect the water from the government bore, after driving the horse and dray to the tank, he would have them stand still while he bucketed the water from the top of the open tank which was just level with the top of his drums.

To do this he would stand on a plank that he laid on top of the drums and with one foot on the plank and one foot on the edge of the tank would transfer the water with his galvanised bucket, 3 gallons at a time.

Usually, in fifteen minutes he would have the drums full and be on his way. Grandma knew that it took him an hour each way with the dray loaded with his drums and allowing fifteen minutes to fill up he would be back in about three and a half hours or thereabouts.

After four hours had passed she became worried, so she saddled her horse and went for a ride to find out what was taking him so long. He had been doing this trip to cart the water for many years without incident and she could not imagine that anything could go wrong.

Her horse was a good galloper so she let the gelding have its head and he fairly stepped out. The tank came into sight after about ten minutes with the horse and dray just away from the tanks with the drums all full.

For a while, she could not see granddad, so she reined in beside the big tank and after dismounting found granddad on the ground where the dray had been.

He was unconscious.

She didn't panic but got a wet cloth and bathed his head and face while she spoke very calmly to him. Shortly he came back to the present. After she asked if he was alright and him telling her that yes he was, thank you, she asked him what had happened to make him fall.

It seemed that just as he was emptying the last bucket of water into the last drum to be filled, a flock of corellas flew into the tree that was just a few yards away from the tank. The screeching of the birds startled the horse and it baulked, moving forward several yards.

With one foot on the tank which did not move and the other on the drum which was now unexpectedly moving, widening the gap between the two, he lost his balance. He tried to twist around and land on the back of the dray but the horses moved a bit faster than they usually did and he fell flat on the ground hitting his head on a boulder that was used to rest the dray wheel against to get it into the right position.

That knocked him out.

He must have been lying there for about an hour. As he was still a bit groggy, he sat there resting against the tank for fifteen minutes or so before he tried to stand. He was a bit unsteady for a few seconds but shortly he began walking around slowly.

Grandma watched as he walked up to the horse and stared squarely into its eyes for a minute or so and then he climbed onto the dray and moved off, heading back to the farm with grandma riding her horse beside him, not really sure if all was right with him yet. He filled the tank for the pigs and emptied the rest of the drums into the horse trough.

He had a sore head for a few days but it didn't seem to slow him down at all.

Uncle Ben had pigs when he was a young man, too. He was known for his prized vegetables at the local Agricultural Show each year and frequently took out some of the top prizes. He married later in life and grandad presented one of his speeches at the wedding reception.

It seems that Uncle Ben was having a few problems with a few pigs that seemed to wander around at night times and get into his vegetable patch and help themselves to some of the prize goodies which grew there. He was not happy about this and reckoned on it being pigs wandering from the next farm. He made sure that his pigs were always in their yard each morning when he went to feed them.

This problem was going on intermittently for some considerable time and Uncle Ben was becoming very frustrated with the loss of his vegetables. At this rate of attrition, he would be lucky to have anything to present at the upcoming agricultural show, let alone provide for his own table.

So one night he set to with his shotgun and he was determined to get those pigs from next door that were destroying his garden.

He perched himself down by the little garden shed that held his tools and potting bench and waited in the dark.

Nothing arrived that night.

So the next night he staked out his position by the little shed once again and just before midnight he watched as a sow entered his veggie patch. It put its nose under the netting wire, lifting it up just enough so that she could crawl under. Once inside the yard,

she took her time wandering around seeking out the best morsel for herself. She passed the pumpkins, went right past the turnips and found herself a nice juicy looking cabbage and began to root around under it. Uncle Ben watched in awe at this big fat sow not only dug up the best and largest cabbage but began eating it as well.

That was just too much for Uncle Ben.

He raised the shotgun to his shoulder, took careful aim and let the old sow have both barrels at the same time.

She died instantly but the cabbage could not be saved, so he threw the partially eaten vegetable into a bucket and he would take this and feed it to his pigs tomorrow.

He left the pig where she was till morning.

After daybreak, he harnessed one of his Clydesdales and with a chain, dragged the old sow to the killing shed. She was so well fed on his vegetables that he was not going to send the carcass back to the neighbour but have it at his own table instead. He butchered the animal before breakfast

and tied a meat bag around it to keep the flies away. After breakfast, he headed off to feed his pigs with some grain and the remains of his prize cabbage.
After feeding the growing pigs he progressed to the dry sow yard, only to find that one of his sows was missing. He carefully scouted around the yard and found the tell-tale marks where she had got out through the fence. He followed the hoof-prints down the creek, along the bank, past the horse yard right up to his vegetable patch.
He had shot his own sow.
There was no newspaper report of the situation that Uncle Ben had found himself in but at church the next Sunday it was all that everyone was talking about.
Uncle Ben was very upset at what had happened but after a while, he could see the funny side to the situation but he would never live that one down.

Modern Stuff

What about new machinery you ask?

Well, this is one thing that really sticks in my mind, as a large leap forward since my family's early days in farming. It seems that the older I get the more changes I see in the modernisation of agriculture.

I am getting on in years and the boys are now running the farm but a few weeks ago, I was sitting in the cab of the son's new tractor as he was seeding this year. It is a new tractor and is fitted with a small seat for the operator to have a trainee beside him so that he could be taught the operation of the tractor during the seeding operation. That's where I park myself to watch the seeding operation and spend some time with the boys and their kids. Sitting on my knee was my grandson, so that made up the three generations. All together in the cabin of the tractor.

It is a big tractor.

The engine is a massive 700 horsepower and it's fitted with a fifteen-speed automatic transmission. There are three tyres, all taller than me, at each corner and the tractor bends in the middle. The cabin is climate controlled so the operator doesn't become fatigued during the day or night. My first tractor, an old Fordson, was like all tractors of that time, an open tractor and after a day's ploughing, I was covered in a heavy coating of brown paddock dust. It gets in your hair, eyes, ears, nose and mouth. Just as well as we had daily showers or else mum would have been really upset with the dirty bedclothes.

There are so many gauges and digital screens around the driver that the cabin looks more like the flight deck of a Boeing 787. The most interesting object is the screen of the GPS unit that is steering the tractor. My son feeds in the coordinates of the paddock and the width of the seeding rig and the machine steers the tractor so he can get the best efficiency out of the thing.

When these auto-steer things were in their infancy stage, there were several situations that I need to tell you about. Some of these have to do with power poles which invade farmer's paddocks. Even the power company doesn't know the exact GPS location of all of their own poles. So when an early unit was set to drive the tractor around the paddock, someone missed a digit of a pole's location.

The entire district's power was out for about a week. It wasn't so much that many poles were brought down by the seeding rig but more to do with the location of the poles which needed to be replaced. It rained heavily after the collision with the pole which snapped off near ground level. The live power wires making sparks all over the ground until the system was shut down after a few minutes.

That was quite a spectacle, one that many will never get to see.

Late that afternoon, the power company sent out a team of about fifteen men and four trucks to replace one power pole but with the rising creek and some flooded ground they were not able to get anywhere near the problem area. For three days they tried to get in to fix the problem but the ruts which are left in the paddock are a testament to the conditions at that time.

Some trees have also been a problem with a large seeding rig like this, so the farmer is faced with a few options. Remove the trees to create a more open paddock *(the greenies didn't like this idea)*, or reprogram the Auto-steer thingy so that it missed the trees. This turned out to be the better option as the greenies didn't bother too much about farm biosecurity back then.

Both options have merit but let's not get sidetracked from what I was trying to tell you.

The larger of the digital screens has about ten segments to it that show the different parts of the seeding rig. There are little cameras at each of the more important points which need to be monitored in case of minor problems. The lad keeps a roving eye on this screen, as well as everything else that is going on. Every couple of hours he needs to get out of the cabin for a change and walks around the rig and sometimes wipes any dust from the camera lenses, as well as a general check of the multitude of bits and pieces.

What's out the back of the tractor is incredible. The first thing I can see is the seeding bar. This is 18 metres wide and has 120 narrow tynes on it so the seed can be dropped into a slot in the ground at the correct spacing and depth. This is carried by 15 rubber wheels and is in hinged segments so that it can be folded up to fit through gateways and so on. Attached to and behind the seeding bar is the seed bin. This holds about 10 tonnes of seed when it's full. The seed bin is fitted with an air blower that sends the seed along some pipes to the bottom of the tynes. Next follows the fertiliser tank with its pump and pipes. This sends the fertiliser along its pipes to the tynes too, so it can be placed under the seed.

I sit and look at this monster as it works the paddock. It travels at about seven kilometres an hour and seeds an area of about 13 hectares each hour it operates. In old terms that is about 30 acres an hour. With his offsider, they operate 24 hours a day and seed over 300 hectares a day. Whew! That's nearly 750 acres a day!

"How much did it cost to set up this outfit?" I asked. He replied with a grin "almost a million dollars."

When my son told me these details my mind went into overdrive and I began remembering things my dad had told me when his father (my grandfather) took up his first land and began farming in the mid-1800s.

His first draught horse cost him 12 pounds with all of the harnesses that were required to hitch the horse to whatever implement that he was connected to. Add that to the 4 pounds that he spent on the single furrow mouldboard plough. His seed bag was cut from an old wheat bag so his first "seeding rig" cost him just 16 pounds.

Did you ask about the difference in production from the early years to modern times? I'm glad I made some notes so that I can give you the right information.

Well, of course, he would have harvested about 6 bags or 18 bushels to the acre in an average year. That relates to about 1 ton per hectare. He would have received about 3 shillings a bushel or 9 shillings per bag or 9 pounds per hectare.

Today my son expects to harvest more than 2 tonnes of wheat per hectare. That should relate to about 550 dollars per hectare. So he'll need a lot of country under crop just to pay for his seeding rig.

But of course, to get a return from his cropping program he'll need to harvest what he's planted and get it delivered to the bin. To complete that part of the job he'll need some more expensive equipment.

My Dad used to cut the crop with a sickle by hand but nowadays with the larger farms, the process is much more expensive. The cost of labour has forced all farmers to invest in larger properties and machinery to make a living from the land.

Nowadays to harvest the crop my son will need a modern header. This machine cuts the standing crop, removes the husk and straw and puts the clean grain into a hopper on the machine.

Well, it's more like a travelling factory really!

From the header's hopper, the grain is transferred into a chaser bin that drives alongside the header so he doesn't have to stop. Sometimes I get the chance to drive this. This, in turn, augers the grain into a storage bin on the farm until a large transport truck can take it to the local bin.

He had one lad working for him a year or so ago who had difficulty in matching the travelling speed of the header when the grain was being augured into the chaser bin that he was pulling behind the open tractor that he was driving. They were in radio contact but he just could not seem to get the speed right. When he drove too fast the grain was spread out behind the chaser bin and wasted on the ground. When he went too slowly, the grain deposited itself on top of him instead of into the bin. Needless to say, he didn't last the season and that's when I had to step up. The wasted grain wasn't too much of a problem when he was running sheep but since he has no sheep to clean up the spilt grain, it's a bit more important to not waste any.

The header is a very complex machine and consists of a cutting bar, spirals, straw walkers, rotating drums, augers, drums, concaves and all sorts of equipment just to complete its part of the process.

To do his job he'll need a large machine with a 9-metre wide front. The new one he bought last year has a 330 horsepower engine, travels at 10 kilometres per hour while harvesting and covers 9 hectares to harvest about 18 tonnes per hour. That machine and his on-farm storage bins cost nearly as much as his seeding rig.

Boy! I'm glad I'm not in the farming game anymore. All that investment to make a living providing food for the nation is unbelievable. And that only works when there is an average or better rainfall year. If the rain falls at the wrong time or not enough of it, or he gets hit by frost, the crop that he harvests won't even pay his interest bill.

But his costs don't stop there. Weeds, insect pests and disease are a big problem now that he doesn't run livestock. He needs a spray rig as well. His old one that he bought about five years ago for 65,000 dollars still does a great job. It has a 20-metre boom and carries 500 litres of liquid for spraying. That's another job he needs to do while the crop is growing. But if the ground is too wet he'll need to have an aircraft do his spraying for him and that is more expensive.

Oh! You may laugh. Did you know that the Australian farmer is one of the biggest gamblers there is?

You gamble that the right amount of rain falls at the right time so the crops can grow as they are supposed to.

You gamble that the machinery doesn't break down, right when it's needed most.

They gamble on the market not falling too much so they can receive a fair price for their product.

They gamble that the government doesn't increase taxes too much. They gamble that the price of fertiliser remains reasonable.

They gamble that the price of fuel stays level.

They gamble that there will be enough workers to handle all of his equipment and that they have sufficient training so they don't bend things too much.

They gamble that when it all goes pear-shaped that the missus stays too.

Boy! What a bunch of gamblers they are!

I've often heard the city folk say that farmers are whingers! Well, I would like to see the average city bloke put up capital and deal with the hardships that we do. More of them would fold than those who have been on the land and are prepared to dig in and do the hard yards when necessary.

Sorry about that! I do tend to get a bit carried away with myself when I talk about these things but I'm sure that you'll understand the difficulties that the farmers face every day.

Well there we are, that's the story in 2020.

What is happening in the future?

What we've seen here is how the descendants of one farming family have lived since the time that they arrived on our shores all that time ago.

But there's more!

Some of our family have been dairy farmers. Some of them have been orchardists. Some of them have been growing potatoes and other vegetables. Some have been sugarcane growers. Some of them have been pastoralists and they will all have a similar story to tell.

But they all would have started off the same way as we have discussed here since page one.

Life on the land can be very tough but those who survive and stay in their industry can tell of so many things that have allowed them to keep their sanity over the years.

What will happen to farming operations in Australia in the future?

The miniaturisation of equipment so that it can run on stored electricity and be automatically recharged by its own solar panels could be an option to keep abreast of costs. Automation of so many pieces of equipment will be so very interesting to watch and be part of.

Automated aerial drones to spot-spray weeds and spray firebreaks and to keep an eye on the crops as they grow. They can even be used to round-up the livestock when they need to be mustered too.

Automatic shearing? That would be very interesting to see, maybe one day....

There are already automatic milking machines, so some of those blokes have a good head start.

I can imagine a seeding unit covered with a solar panel automatically seeding by itself and not needing any human contact. The larger the cropping program the more units that are needed. I can also see one of these units going astray and trying to plant a crop around Parliament House in Canberra. It may not need any fertiliser as there is always enough being spread around there all the time. Ha-ha-ha.

But, seriously, we could be producing some of our foodstuffs on the Moon or even Mars in twenty or thirty years.

Well, it seems that that is my story told, from way back when our family arrived, right up to the present day.

So much change has occurred in the past and if things keep changing exponentially, the way it has in the past, who knows what the future holds in twenty years.

Just imagine if someone picks up this story in the year 2050 and compares the details of our present time to their present time, how much change there will be. Much of that change is unimaginable just now.

One thing is for sure though, there will always be the need to produce food.

About the Author

David Kentish spent his early years on the family dairy farm just south of Perth in Western Australia near the small settlement of Keysbrook.

Before the time of broadcast television, his father, J. Lance Kentish, spent time in the evenings inventing and telling stories about the bush animals, the talking red-gum tree and the magic carpet to his family.

David has continued in this same vein with the telling of stories of imaginary Australian bush animals and friends and the many predicaments that they find themselves involved in.

He and his wife Barbara enjoy travelling with their 4x4 and caravan in and around the Australian outback and bush. This is where he gets most of his inspiration which has led to a collection of stories of their travels.

The Kentish's of Keysbrook, tells the story of his family and the trials and tribulations of their lives as farmers from 1838 onwards.

King's Gold tells of an adventure that went wrong and how several people ended up doing something very unexpected. Set in some of the most picturesque outback country in Western Australia.

A Place Called Earth was written whilst on one of these trips and is an exciting collection of stories from David's point of view of how the earth began and developed over many countless millennia. A well-written yarn that will keep you intrigued right up to the last page.

Beside the Billabong, tells the story of five differing bush friends who learn to work together and help each other survive some very harsh conditions beside a billabong in outback Australia.

KVK, *A life well lived;* is a biography of his mother and the struggles that she endured as a youngster through her life until she departed this Earth.

The Little Pink Dragon Who Couldn't is a yarn about someone who didn't fit in and wasn't able to perform as her peers. Depression took hold of her life so she set out to find a remedy for herself. A story with a successful ending.

David has several more stories in the pipeline so keep a look-out for more stories by David Kentish.

Other books by David Kentish

Visit his website for more detail
www.davidkentish.com.au

www.ingramcontent.com/pod-product-compliance
Lightning Source LLC
Chambersburg PA
CBHW050312010526
44107CB00055B/2215